SpringerBriefs in Physics

SpringerBriefs in Physics are a series of slim high-quality publications encompassing the entire spectrum of physics. Manuscripts for SpringerBriefs in Physics will be evaluated by Springer and by members of the Editorial Board. Proposals and other communication should be sent to your Publishing Editors at Springer.

Featuring compact volumes of 50 to 125 pages (approximately 20,000-45,000 words), Briefs are shorter than a conventional book but longer than a journal article. Thus, Briefs serve as timely, concise tools for students, researchers, and professionals.

Typical texts for publication might include:

- A snapshot review of the current state of a hot or emerging field
- A concise introduction to core concepts that students must understand in order to make independent contributions
- An extended research report giving more details and discussion than is possible in a conventional journal article
- A manual describing underlying principles and best practices for an experimental technique
- An essay exploring new ideas within physics, related philosophical issues, or broader topics such as science and society

Briefs allow authors to present their ideas and readers to absorb them with minimal time investment.

Briefs will be published as part of Springer's eBook collection, with millions of users worldwide. In addition, they will be available, just like other books, for individual print and electronic purchase.

Briefs are characterized by fast, global electronic dissemination, straightforward publishing agreements, easy-to-use manuscript preparation and formatting guidelines, and expedited production schedules. We aim for publication 8-12 weeks after acceptance.

More information about this series at http://www.springer.com/series/8902

Sunkyu Yu · Xianji Piao · Namkyoo Park

Top-Down Design of Disordered Photonic Structures

Multidisciplinary Approaches Inspired by Quantum and Network Concepts

Sunkyu Yu
Department of Electrical
and Computer Engineering
Seoul National University
Seoul, Korea (Republic of)

Xianji Piao
Department of Electrical
and Computer Engineering
Seoul National University
Seoul, Korea (Republic of)

Namkyoo Park
Department of Electrical
and Computer Engineering
Seoul National University
Seoul, Korea (Republic of)

ISSN 2191-5423   ISSN 2191-5431   (electronic)
SpringerBriefs in Physics
ISBN 978-981-13-7526-2   ISBN 978-981-13-7527-9   (eBook)
https://doi.org/10.1007/978-981-13-7527-9

This Springer imprint is published by the registered company Springer Nature Singapore Pte Ltd.
The registered company address is: 152 Beach Road, #21-01/04 Gateway East, Singapore 189721, Singapore

Acknowledgements

S. Yu was supported by the Presidential Postdoctoral Fellowship of the Basic Science Research Program (2016R1A6A3A04009723) funded by the Korean government through the National Research Foundation of Korea (NRF). X. Piao and N. Park were supported by the Korea Research Fellowship (KRF, 2016H1D3A1938069) Program through the NRF. This work was also supported by the Global Frontier Program (GFP, 2014M3A6B3063708) through the NRF.

Contents

Chapter 1
Introduction

Abstract This chapter introduces the concept and motivation of disordered photonics. Starting from the definitions of order, disorder, and randomness, we describe the importance of exploring the intermediate areas of disorder between order and randomness in physics. With a brief summary of the role of order and randomness in photonics, we introduce a top-down implementation of anomalous photonic disorder as a toolkit for the independent control of optical quantities.

Keywords Order · Disorder · Random · Small world · Top-down design · Eigensystem

1.1 Order, Randomness, and Their Intermediate Area

Because of their similar etymological roots, the terminologies "order" and "disorder" have often been misunderstood as being opposite concepts. However, let us take an example in physics. Order is a definite and mathematically rigorous concept: a crystal is a lattice arrangement of atoms with discrete translational symmetries (Marder and Wiley 2010), and a quasicrystal, such as the Fibonacci sequence or Penrose tiling, is generated by the cut-and-projection method applied to higher-dimensional crystals (Levine and Steinhardt 1984). Other types of aperiodic order, such as Thue-Morse (Cheng et al. 1988) or Rudin–Shapiro (Dulea et al. 1992) sequences, are defined by rigorous generating rules. Compared with these rigorous definitions of order, how can we define disorder in physics?

First, we need to distinguish "disorder" from "randomness". The prefix *dis-*, meaning "*not*", shows that "dis"-order covers whole regimes of partially or completely broken order, in contrast to randomness, which represents the *complete* lack of *any* order. Contrary to the rigorous definitions of order and the statistically strict definition of randomness, "disorder" thus has inherent ambiguity in relation to the *strength* and *type* of the breaking of order. While order and randomness are specialized and extreme "subsets" of disorder, there are infinite subsets of disorder by combining order and randomness in different ways.

© The Author(s), under exclusive license to Springer Nature Singapore Pte. Ltd. 2019
S. Yu et al., *Top-Down Design of Disordered Photonic Structures*,
SpringerBriefs in Physics, https://doi.org/10.1007/978-981-13-7527-9_1

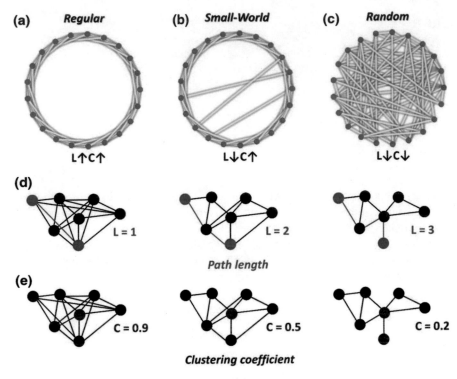

Fig. 1.1 Small-world network. **a** Regular, **b** small-world, and **c** random graphs, defined by two mathematical indicators: **d** path length L and **e** clustering coefficient C. Each circle denotes a "node", and each line denotes a "link". The blue and red circles (or lines) in **d** and **e** represent the example nodes (or links) for the values shown for L and C, respectively. For example, in the right figure of **d**, blue nodes are connected through three links (L = 3) in the shortest path. In the middle figure of **e**, the red node has five neighbour nodes, and thus, the number of possible connections between the nodes is $5 \times 4/2 = 10$. Because five connections exist between neighbouring nodes, the clustering coefficient is C = 5/10 = 0.5. The overall characteristics of the graph are determined by the averages of L and C for all nodes and links. The graphs are obtained by the WS rewiring process (Watts and Strogatz 1998)

Why do we need to consider these ambiguous and mathematically unsettled intermediate areas between order and randomness? How can we quantify useful regimes between order and randomness? There is a convincing example in network theory (Fig. 1.1), as demonstrated in the pioneering work by Watts and Strogatz (1998). Consider order and randomness in graph networks: regular (Fig. 1.1a) and random (Fig. 1.1c) graphs, respectively. **The properties of graphs** composed of nodes and links are quantified by two indicators **(P- for property)**:

P-I. Path length L (Fig. 1.1d): **Typical separation between two nodes** in a graph.
P-II. Clustering coefficient C (Fig. 1.1e): **Proportion of realized connections** among neighbours compared with the number of all possible connections.

From its definition, path length L determines the efficiency of signal transport inside a system, with a smaller L for faster transport. On the other hand, the clustering coefficient C determines the robustness of the system, with a higher C for more robust systems, because more links need to be removed to prevent signal transport inside systems. With these statistical indicators, regular graphs lead to inefficient but stable signal transport (large L and high C), while random graphs allow for efficient but fragile signal transport (small L and low C).

Between these ordered and random graphs, D. J. Watts discovered a novel "small-world" regime with a small L and high C (Fig. 1.1b) (Watts and Strogatz 1998). Because of the characteristics defined by L and C, the signal transport inside the small-world graph is distinct from both regular and random graphs, with more efficient signal transport than that of the regular graph and more robust signal transport than that of the random graph (there is also the issue of the rare but dramatic degradation of the signal transport efficiency in small-world networks, i.e., the removal of the "hub" in scale-free networks, e.g., see Xia et al. 2010). From this originality and the advantages of the small-world graph, we can find examples of small-world structures in various fields including biological systems, such as neurons of *Caenorhabditis elegans* and brain connectomes; social and commercial systems, such as affinity groups and world airport networks; and seismic networks in geophysics.

The key concept of the small world is the different sensitivities of the two indicators L and C to randomization. When we increase the randomness in the regular graph (e.g., following the Watts-Strogatz (WS) rewiring process in Watts and Strogatz 1998), the value of L decreases more rapidly than that of C, resulting in an exotic intermediate regime distinct from regular and random networks. This result shows important **viewpoints on the handling of intermediate areas between order and randomness (V-for viewpoint)**.

V-I. Use of "multiple" key parameters to quantify the characteristics of the system illuminates the intermediate areas between order and randomness.
V-II. Do these quantities change "identically" with randomization? If not, there may be an "anomalous" type of disorder distinct from both order and randomness.

From this viewpoint, we introduce the field of disordered photonics (Wiersma 2013): the analysis of light-matter interactions in disordered structures and the design of disordered structures for target photonic functionalities. In particular, we focus on finding and using "anomalous" types of disorder in the intermediate areas between order and randomness.

1.2 Order, Randomness, and Anomalous Disorder in Photonics

In photonics, material composition with respect to order and randomness has been one of the most important characteristics not only for understanding wave phenomena but also for establishing the purpose of photonic devices. The basic theories and applications of classical optics have usually been developed with homogeneous, linear, and static media (Teich and Saleh 2007; Yariv and Yeh 2006), corresponding to the ultimate "order" that satisfies most symmetries for electromagnetics, including translational, rotational, mirror, and time-reversal symmetries and reciprocity. A slightly broken order has been studied to manipulate the flow of light, including crystal optics with discrete translational symmetry (Joannopoulos et al. 2011; Thiel et al. 2009; Chung et al. 2012; Woo et al. 2012), basic optical elements constructed by including defects (Joannopoulos et al. 2011; Okamoto 2010; Lee et al. 2012; Mason et al. 2014; Kim et al. 2017; Notomi et al. 2008; Nozaki et al. 2010), anharmonic potentials for nonlinear optics (Boyd and Elsevier 2003; Yi et al. 2015; Yoo et al. 2017, 2019), and time-varying media for broken time-reversal symmetry (Yu and Fan 2009; Tanaka et al. 2007; Sounas and Alù 2017; Huang et al. 2018). In terms of light-matter interactions, the structural order in materials such as gratings, photonic crystals, and fractal structures derives the conservation of the wave quantities (e.g., wavevector, frequency, and angular momentum) protected by symmetries (e.g., translational, rotational, and chiral). These conserved quantities lead to **two representative properties of photonic order (O-for order):**

O-I. Spatially extended modes defined by Bloch's theorem (Fig. 1.2a).
O-II. Perfect bandgaps from coherent scattering and the resulting interference (Fig. 1.2b).

The purpose of photonic devices based on ordered structures has thus been focused on signal transport and the control of signal transport induced by different types of symmetry breaking, such as guiding (Snyder and Love 2012; Slussarenko et al. 2016; Bozhevolnyi et al. 2006), bending (Gabrielli et al. 2012), beaming (Li et al. 2015; Kumar et al. 2010a, b), and lensing (Aieta et al. 2012; Chen et al. 2013, 2018; Jung et al. 2009).

Instead, optical phenomena in random structures, such as randomly inhomogeneous refractive index profiles and randomly coupled optical elements, result in incoherent and randomized wave behaviours. These chaotic wave phenomena with interference determine **the core characteristics of light-matter interactions in random photonic structures (R-for random):**

R-I. Emergence of Anderson localization (Anderson 1958; Schwartz et al. 2007), or spatially localized modes, from the interference between random scattering (Fig. 1.2c).
R-II. Annihilation of bandgaps (Ashcroft and Mermin 1976; Li and Zhang 2000) or broadband responses (Bozzola et al. 2014) (Fig. 1.2d).

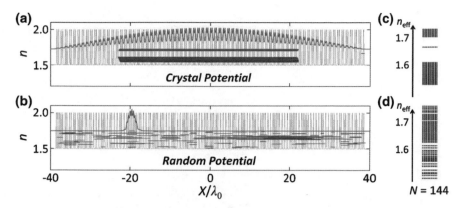

Fig. 1.2 Order and randomness in photonics. **a, b** Crystal and **c, d** Bernoulli random potentials: the corresponding eigensystem: **a, c** eigenmodes, and **b, d** eigenvalues (or effective indices n_{eff}). Black lines indicate the refractive index potential landscapes. Blue curves in **a** and **c** denote the ground state. The x-axis length and y-axis position of each straight line in **a** and **c** represent the modal size and eigenvalue of each eigenmode. Spatially extended eigenmodes (O-I) and strong modal localization (R-I) can be found in **a** and **c**, respectively. Additionally, emergence (O-II) and annihilation (R-II) of the bandgap can be found in **b** and **d**, respectively. Figure and caption adapted from (Yu et al. 2015) under a CC BY license (http://creativecommons.org/licenses/by/4.0/)

For the realization of photonic devices, random structures have been exploited in a restricted manner, such as the focusing of optical energy from Anderson localization and the design of broadband devices without bandgap phenomena, which are inherently distinguished from photonic transporting devices based on ordered structures.

From the properties of order and randomness in photonics, we can now raise two questions (**Q-**) and answers (**A-**) based on the discussion in Sect. 1.1:

Q-I. What are the "multiple" key parameters to quantify the characteristics of order and randomness in light-matter interactions?
A-I. The modal (O-I vs. R-I) and spectral (O-II vs. R-II) features successfully define the characteristics of order and randomness.
Q-II. Do these quantities change "identically" with randomization? If not, there may be an "anomalous" type of disorder distinct from both order and randomness.
A-II. We show in Chaps. 2 and 3 that these quantities can be independently controlled, allowing top-down realization of anomalous photonic disorder distinct from both order and randomness: (i) perfect bandgaps with localization (**O-II** *and* **R-I, Chap.** 2) and (ii) wave transport with random-like spectral information with robustness (**O-I** *and* **R-II, Chap.** 3).

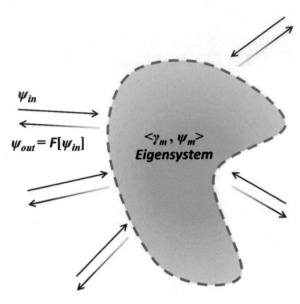

Fig. 1.3 Understanding light-matter interactions in terms of eigensystems: sets of eigenmodes and the corresponding eigenvalues. The relationship between input (ψ_{in}) and output (ψ_{out}) is defined by an abstract scattering function F, which has the parameters of the eigensystem $<\gamma_m, \psi_m>$ and boundary conditions

1.3 Top-Down Design Based on Eigensystems

In Chaps. 2 and 3, we introduce our recent achievements (Yu et al. 2015, 2016a, b, 2017a, b, 2018a, b) in the deterministic design of optical phenomena, overcoming the current limitations in highly disordered optical structures. The underlying concept of our approach is to understand light-matter interactions in terms of the eigensystem (Fig. 1.3): the set of all eigenmodes of a linear operator, each paired with its corresponding eigenvalue. We develop top-down design methodologies of disordered photonic structures by independently controlling eigenvalues (or spectral properties) or eigenmodes (or modal properties) through inverse engineering of the governing wave equation, designing photonic disorder from target wave quantities.

We also develop new concepts for disordered photonics inspired by concepts in quantum mechanics, solid-state physics, mathematics, and network theory, e.g., isospectrality, supersymmetry, graph networks, small-world, de Broglie-Bohm theory, and parity-time symmetry. This multidisciplinary approach provides a new perspective on the design methodology in photonics, especially in terms of the top-down design of future devices in photonics, including robust photonic bandgaps and wave dynamics, broadband switching, fuzzy logics, and a new way of optical energy storage and phase trapping from Bohmian photonics. This book provides new design strategies for physicists and engineers in photonics and shows the inspirations of researchers in other fields from multidisciplinary approaches.

References

Aieta, F., Genevet, P., Kats, M.A., Yu, N., Blanchard, R., Gaburro, Z., Capasso, F.: Aberration-free ultrathin flat lenses and axicons at telecom wavelengths based on plasmonic metasurfaces. Nano Lett. **12**, 4932–4936 (2012)

Anderson, P.W.: Absence of diffusion in certain random lattices. Phys. Rev. **109**, 1492 (1958)

Ashcroft, N.W., Mermin, N.D., Rodriguez, S.: Solid state physics. Cengage Learn. (1976)

Boyd, R.W.: Nonlinear Optics. Elsevier (2003)

Bozhevolnyi, S.I., Volkov, V.S., Devaux, E., Laluet, J.Y., Ebbesen, T.W.: Channel plasmon subwavelength waveguide components including interferometers and ring resonators. Nature **440**, 508–511 (2006). https://doi.org/10.1038/nature04594

Bozzola, A., Liscidini, M., Andreani, L.C.: Broadband light trapping with disordered photonic structures in thin-film silicon solar cells. Prog. Photovolt: Res. Appl. **22**, 1237–1245 (2014)

Cheng, Z., Savit, R., Merlin, R.: Structure and electronic properties of Thue-Morse lattices. Phys. Rev. B **37**, 4375 (1988)

Chen, X., Park, H.-R., Pelton, M., Piao, X., Lindquist, N.C., Im, H., Kim, Y.J., Ahn, J.S., Ahn, K.J., Park, N.: Atomic layer lithography of wafer-scale nanogap arrays for extreme confinement of electromagnetic waves. Nat. Commun. **4**, 2361 (2013)

Chen, W.T., Zhu, A.Y., Sanjeev, V., Khorasaninejad, M., Shi, Z., Lee, E., Capasso, F.: A broadband achromatic metalens for focusing and imaging in the visible. Nat. Nanotech. **13**, 220 (2018)

Chung, K., Yu, S., Heo, C.J., Shim, J.W., Yang, S.M., Han, M.G., Lee, H.S., Jin, Y., Lee, S.Y., Park, N.: Flexible, angle-independent, structural color reflectors inspired by morpho butterfly wings. Adv. Mater. **24**, 2375–2379 (2012)

Dulea, M., Johansson, M., Riklund, R.: Trace-map invariant and zero-energy states of the tight-binding Rudin-Shapiro model. Phys. Rev. B **46**, 3296 (1992)

Gabrielli, L.H., Liu, D., Johnson, S.G., Lipson, M.: On-chip transformation optics for multimode waveguide bends. Nat. Commun. **3**, 1217 (2012)

Huang, R., Miranowicz, A., Liao, J.-Q., Nori, F., Jing, H.: Nonreciprocal photon blockade. Phys. Rev. Lett. **121**, 153601 (2018)

Joannopoulos, J.D., Johnson, S.G., Winn, J.N., Meade, R.D.: Photonic Crystals: Molding The Flow of Light. Princeton University Press (2011)

Jung, Y.J., Park, D., Koo, S., Yu, S., Park, N.: Metal slit array Fresnel lens for wavelength-scale optical coupling to nanophotonic waveguides. Opt. Express **17**, 18852–18857 (2009)

Kim, Y., Yu, S., Park, N.: Low-dimensional gap plasmons for enhanced light-graphene interactions. Sci. Rep. **7**, 43333 (2017)

Kumar, M.S., Piao, X., Koo, S., Yu, S., Park, N.: Out of plane mode conversion and manipulation of Surface Plasmon Polariton Waves. Opt. Express **18**, 8800–8805 (2010a)

Kumar, M.S., Menabde, S., Yu, S., Park, N.: Directional emission from photonic crystal waveguide terminations using particle swarm optimization. JOSA B **27**, 343–349 (2010b)

Lee, H., Chen, T., Li, J., Yang, K.Y., Jeon, S., Painter, O., Vahala, K.J.: Chemically etched ultrahigh-Q wedge-resonator on a silicon chip. Nat. Photon. **6**, 369–373 (2012)

Levine, D., Steinhardt, P.J.: Quasicrystals: a new class of ordered structures. Phys. Rev. Lett. **53**, 2477 (1984)

Li, Z.-Y., Zhang, Z.-Q.: Fragility of photonic band gaps in inverse-opal photonic crystals. Phys. Rev. B **62**, 1516 (2000)

Li, L., Li, T., Tang, X.-M., Wang, S.-M., Wang, Q.-J., Zhu, S.-N.: Plasmonic polarization generator in well-routed beaming. Light Sci. Appl. **4**, e330 (2015)

Marder, M.P.: Condensed Matter Physics. Wiley (2010)

Mason, D.R., Menabde, S.G., Yu, S., Park, N.: Plasmonic excitations of 1D metal-dielectric interfaces in 2D systems: 1D surface plasmon polaritons. Sci. Rep. **4**, 4536 (2014)

Notomi, M., Kuramochi, E., Tanabe, T.: Large-scale arrays of ultrahigh-Q coupled nanocavities. Nat. Photon. **2**, 741–747 (2008). https://doi.org/10.1038/nphoton.2008.226

Nozaki, K., Tanabe, T., Shinya, A., Matsuo, S., Sato, T., Taniyama, H., Notomi, M.: Sub-femtojoule all-optical switching using a photonic-crystal nanocavity. Nat. Photon. **4**, 477–483 (2010)

Okamoto, K.: Fundamentals of Optical Waveguides. Academic Press (2010)

Schwartz, T., Bartal, G., Fishman, S., Segev, M.: Transport and Anderson localization in disordered two-dimensional photonic lattices. Nature **446**, 52–55 (2007)

Slussarenko, S., Alberucci, A., Jisha, C.P., Piccirillo, B., Santamato, E., Assanto, G., Marrucci, L.: Guiding light via geometric phases. Nat. Photon. **10**, 571 (2016)

Snyder, A.W., Love, J.: Optical Waveguide Theory. Springer Science & Business Media, Boston (2012)

Sounas, D.L., Alù, A.: Non-reciprocal photonics based on time modulation. Nat. Photon. **11**, 774 (2017)

Tanaka, Y., Upham, J., Nagashima, T., Sugiya, T., Asano, T., Noda, S.: Dynamic control of the Q factor in a photonic crystal nanocavity. Nat. Mater. **6**, 862–865 (2007). https://doi.org/10.1038/nmat1994

Teich, M.C., Saleh, B.: Fundamentals of Photonics. Wiley Interscience (2007)

Thiel, M., Rill, M.S., von Freymann, G., Wegener, M.: Three-dimensional Bi-Chiral photonic crystals. Adv. Mater. **21**, 4680–4682 (2009)

Watts, D.J., Strogatz, S.H.: Collective dynamics of 'small-world' networks. Nature **393**, 440–442 (1998)

Wiersma, D.S.: Disordered photonics. Nat. Photon. **7**, 188–196 (2013)

Woo, I., Yu, S., Lee, J.S., Shin, J.H., Jung, M., Park, N.: Plasmonic structural-color thin film with a wide reception angle and strong retro-reflectivity. Photonics J., IEEE **4**, 2182–2188 (2012)

Xia, Y., Fan, J., Hill, D.: Cascading failure in Watts-Strogatz small-world networks. Phys. A **389**, 1281–1285 (2010)

Yariv, A., Yeh, P.: Photonics: Optical Electronics in Modern Communications. Oxford University Press, Inc. (2006)

Yi, J.-M., Smirnov, V., Piao, X., Hong, J., Kollmann, H., Silies, M., Wang, W., Groß, P., Vogelgesang, R., Park, N.: Suppression of radiative damping and enhancement of second harmonic generation in bull's eye nanoresonators. ACS Nano **10**, 475–483 (2015)

Yoo, K., Becker, S.F., Silies, M., Yu, S., Lienau, C., Park, N.: Bridging Microscopic Nonlinear Polarizations Toward Far-Field Second Harmonic Radiation (2017). arXiv:1711.09568

Yoo, K., Becker, S.F., Silies, M., Yu, S., Lienau, C., Park, N.: Steering second-harmonic radiation through local excitations of plasmon. Opt. Express **27**, 18246 (2019)

Yu, Z., Fan, S.: Complete optical isolation created by indirect interband photonic transitions. Nat. Photon. **3**, 91–94 (2009). https://doi.org/10.1038/nphoton.2008.273

Yu, S., Piao, X., Hong, J., Park, N.: Bloch-like waves in random-walk potentials based on supersymmetry. Nat. Commun. **6**, 8269 (2015)

Yu, S., Piao, X., Hong, J., Park, N.: Metadisorder for designer light in random systems. Sci. Adv. **2**, e1501851 (2016a)

Yu, S., Piao, X., Hong, J., Park, N.: Interdimensional optical isospectrality inspired by graph networks. Optica **3**, 836–839 (2016b)

Yu, S., Piao, X., Park, N.: Target decoupling in coupled systems resistant to random perturbation. Sci. Rep. **7**, 2139 (2017a)

Yu, S., Piao, X., Park, N.: Controlling random waves with digital building blocks based on supersymmetry. Phys. Rev. Appl. **8**, 054010 (2017b)

Yu, S., Piao, X., Park, N.: Disordered potential landscapes for anomalous delocalization and superdiffusion of light. ACS Photon. **5**, 1499 (2018a)

Yu, S., Piao, X., Park, N.: Bohmian photonics for independent control of the phase and amplitude of waves. Phys. Rev. Lett. **120**, 193902 (2018b)

Chapter 2
Designing Spectra in Disordered Photonic Structures

Abstract In this chapter, we introduce the inverse design of disordered photonic structures with target spectral information. Starting from the concept of isospectrality, which originated from the old question of *"Can one hear the shape of a drum?"*, we discuss the derivation of isospectral relations between different shapes of photonic structures. Two mathematical techniques are introduced as examples of the nontrivial isospectral transformation in disordered photonics, supersymmetry and the Householder transformation, which enable the practical applications of photonic bandgap media with strong modal localization, random wave switching for binary or fuzzy logics, and inter-dimensional light transport without any loss of information.

Keywords Isospectrality · Supersymmetry · Photonic bandgap · Random wave switching · Householder transformation · Inter-dimensional light transport

2.1 Concept of Isospectrality

"Can one hear the shape of a drum?" (Kac 1966) This old and elegant question was raised by Mark Kac in American Mathematical Monthly in 1966. As we all know, drums make different sounds according to their shapes. Depending on the shape, or more precisely, the elastic potential landscape of the drum membrane, there exists a set of allowed tones for the sound defined by the eigenvalues of the elastic membrane potential. The sound of the drum is then determined by the linear combination of these allowed tones. Therefore, if we are able to "hear" the shape of a drum, the drum shape should be uniquely determined by the allowed tones of the drum sound, which requires one-to-one correspondence between a set of eigenvalues and the potential landscape. In contrast, the existence of different shapes of drums that generate the same sound spectrum prohibits the exact prediction of the drum shape (Fig. 2.1). Notably, this mathematical interpretation has generalized the question to other physical domains: "Can one hear the shape of a graph?" (Gutkin and Smilansky 2001), "Can one hear the thermodynamics of a (rough) colloid?" (Duplantier 1991), and "Can one hear the dimension of a fractal?" (Brossard and Carmona 1986).

© The Author(s), under exclusive license to Springer Nature Singapore Pte. Ltd. 2019
S. Yu et al., *Top-Down Design of Disordered Photonic Structures*,
SpringerBriefs in Physics, https://doi.org/10.1007/978-981-13-7527-9_2

Fig. 2.1 Rabbit is trying to predict the shape of the veiled drum. Is it possible to uniquely predict the drum shape with only the sound spectrum?

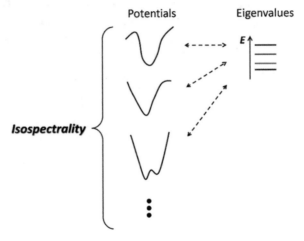

Fig. 2.2 Set of isospectral potentials with identical eigenvalue spectral information. Different potential landscapes can derive identical eigenvalues from the concept of isospectrality. The finding of the isospectral potentials is based on the isospectral transformation discussed in Sects. 2.2–2.6

The answer to that historical question is no, as revealed in Gordon et al. (1992), "one cannot hear the shape of a drum". Because different shapes of elastic potentials can exist with identical eigenvalue spectra, we cannot uniquely predict the shape of a drum with a given sound spectrum. This relationship is defined by the concept of "isospectrality" between potentials (Fig. 2.2): the relationship between potentials with identical (or *iso-*) spectral information. Isospectrality between different elastic potentials has been extended to other fields (Gutkin and Smilansky 2001; Duplantier 1991; Brossard and Carmona 1986) with various applications, such as quantum phase detections (Moon et al. 2008) and the modelling of anyons (Keilmann et al. 2011).

How can we obtain the isospectral relations between different shapes of potentials? To illustrate the derivation of isospectral relations in various fields, consider the general Hamiltonian eigenvalue equation $H\psi_m = \gamma_m \psi_m$ ($m = 0, 1, \ldots$), which describes wave behaviours in quantum mechanics (Sakurai and Tuan 1995), photonics (Longhi 2009), electronics (Schindler et al. 2011), and acoustics (Zhu et al. 2014).

If we assume the application of a certain transformation operator A to the Hamiltonian equation, the equation becomes $AH\psi_m = \gamma_m(A\psi_m)$. Because the isospectrality involves the conservation of the eigenvalue γ_m, we develop the "intertwining relation" (Kuru et al. 2001; Demircioğlu et al. 2002) of $AH = H_{iso}A$, which leads to the complete isospectral relation of $H_{iso}(A\psi_m) = \gamma_m(A\psi_m)$ when $A\psi_m \neq O$ for all m. The operator A is then called the "intertwining operator" (Kuru et al. 2001; Demircioğlu et al. 2002). If H_{iso} and $A\psi_m$ are not the results of the *trivial* transformations (e.g., translation and rotation) of H and ψ_m, respectively, the potentials included in both H and H_{iso} are inherently different and thus satisfy the *nontrivial* isospectral relation with (i) the identical eigenspectrum defined by γ_m and (ii) the transformed eigenmode from ψ_m to $A\psi_m$. We therefore summarize **two key properties of isospectral transformations (P- for property)**:

P-I. Conserved eigenspectrum γ_m $(m = 0, 1, ...)$ and
P-II. Transformed eigenmode $(\psi_m \rightarrow A\psi_m)$ with the transformed potential (or Hamiltonian $AH = H_{iso}A$),

which are obtained from **the sufficient conditions of the isospectrality (C- for condition)**:

C-I. $AH = H_{iso}A$, and
C-II. $A\psi_m \neq O$ **for all** m $(m = 0, 1, ...)$.

These properties and conditions in the isospectral relation allow independent target control of the eigenvalues and eigenmodes in physical systems, as shown in Sects. 2.2–2.6.

Then, what can we obtain in disordered photonics using the concept of isospectrality? For example, can we "see" the shape of optical potentials (with only a spectrum of light)? As with the case of the drum, the answer is no, we cannot predict the exact potential landscape with only the spectral information of scattered light. We can therefore envisage the existence of different shapes of photonic structures that have the same spectral response but different modal properties. This photonic isospectrality enables the separation of photonic quantities especially in the intermediate regime between order and randomness as discussed in Sects. 1.1–1.2. For example, if one cannot "predict" the crystalline structure with only the bandgap "spectrum" of light, can we discover a disordered photonic structure that has an isospectral relation with a crystal? Can we achieve a disordered potential landscape simultaneously possessing crystal-like spectral information (e.g., perfect bandgap) and random-like modal properties (e.g., Anderson localization)? Can we design disordered potentials with the same spectral information and totally different modal properties (e.g., mode profiles with different dimensions)? As answers to these questions, we introduce two different types of nontrivial isospectral transformations and their applications: supersymmetric transformation (Sects. 2.2–2.4) and Householder transformation (Sects. 2.5–2.6).

2.2 Isospectrality by Supersymmetric Transformation

The field of supersymmetry (Ramond 1971) (SUSY) shares various properties with the concept of isospectrality. SUSY is one of the most important postulates introduced for a unified description of basic interactions in physics (Cooper et al. 2001) by defining the relationship between bosons and fermions (Fig. 2.3). Over the past few decades, SUSY has been considered a promising postulate, especially in particle physics, for completing the Standard Model in terms of string theory (Ramond 1971). Because unbroken SUSY predicts the existence of the SUSY partners of quarks and leptons with the same masses as their SUSY counterparts, the lack of such particles may represent spontaneous breaking of SUSY in nature (Witten 1981).

Although the experimental evidence of this ambitious postulate is still lacking (Ellis 2015) and has encountered serious controversy (Woit 2011; Lykken and Spiropulu 2014) in high-energy physics, the concept of SUSY and its elegant mathematical formulations related to isospectrality have provided remarkable opportunities in various fields, e.g., (i) SUSY quantum mechanics (Cooper et al 2001; Bagrov and Samsonov 1995; Schmidt and Friedrich 2014), (ii) SUSY photonics (Miri et al. 2013a, b; Heinrich et al. 2014a, b; Yu et al. 2015a, b; 2017, 2018; Hokmabadi et al. 2019; Longhi 2015; Barkhofen et al. 2018; Miri et al. 2013a, b, 2014; Midya et al. 2018; Smirnova et al. 2019; Queraltó et al. 2018; Teimourpour et al. 2016; Principe et al. 2015; Chumakov and Wolf 1994; Laba and Tkachuk 2014; Heinrich et al. 2014a, b; Walasik et al. 2018; Macho et al. 2018; El-Ganainy et al. 2015; Zuniga-Segundo et al. 2014; Longhi 2014), (iii) random field Ising model (Tissier and Tarjus 2011), (iv) graph theory (Nakata et al. 2016), (v) mechanical systems (Chaunsali et al. 2017), and (vi) topological modes (Grover et al. 2014). In this book, we focus only on the "isospectral properties of the SUSY transformation" in SUSY photonics. First, we introduce the basic formulation of the SUSY with the Hamiltonian equation, which has been intensively studied in quantum mechanics and recently in photonics.

Hamiltonian Factorization for SUSY The SUSY relationship is mathematically defined by the Darboux transformation (Bagrov and Samsonov 1995). Consider again the Hamiltonian eigenvalue equation $H\psi_m = \gamma_m\psi_m$ ($m = 0, 1, \ldots$). If we

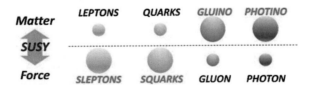

Fig. 2.3 SUSY relationships between fermions (matter particles) and bosons (force particles). The unbroken SUSY postulate predicts the existence of paired relationships between bosons and fermions at the same mass level

assume that the Hamiltonian can be decomposed into two partial operators as $H = BA + \gamma_d$ for the Darboux transformation (Bagrov and Samsonov 1995), where γ_d is a scalar constant, the equation can be rewritten as

$$(BA + \gamma_d)\psi_m = \gamma_m \psi_m. \tag{2.1}$$

When we apply the partial operator A to Eq. (2.1), we achieve the transformed equation

$$(AB + \gamma_d)(A\psi_m) = \gamma_m(A\psi_m), \tag{2.2}$$

which automatically satisfies the isospectrality with the transformed eigenmode $A\psi_m$. Therefore, the factorization of the Hamiltonian $H = BA + \gamma_d$ allows the construction of the isospectral Hamiltonian $H_{iso} = AB + \gamma_d$, which satisfies one of the sufficient conditions of the isospectrality, the intertwining relationship $AH = H_{iso}A = ABA + \gamma_d A$. Therefore, the partial operator A corresponds to the intertwining operator.

However, the factorization of H does not always guarantee complete isospectrality, because the transformation by the intertwining operator A could lead to $A\psi_n = O$ for the nth eigenmode ($n \in \{m\}$). This "annihilation" of the transformed mode leads to the absence of the nth eigenvalue in the eigenspectrum of the transformed Hamiltonian H_{iso}, prohibiting the complete isospectral relation, rather being close to the *quasi*-isospectrality. Notably, from Eq. (2.2), the condition of $\gamma_d = \gamma_n$ leads to the relation $BA\psi_n = O$, which is the necessary condition of the annihilation $A\psi_n = O$ and the breaking of the complete isospectrality. This property is one of the key properties of the SUSY transformation, the classification of the unbroken and broken SUSY, which is discussed later.

SUSY in Schrödinger-like Hamiltonian For the factorization of the Hamiltonian H, we focus on the 1-dimensional Schrödinger-like equation (Longhi 2009; Miri et al. 2013a, b; Yu et al. 2015a, b; Dreisow et al. 2009). Because this basic formulation is valid for both a nonrelativistic particle in quantum mechanics and a transverse electric (TE) optical mode, we can explore the SUSY-induced isospectrality in quantum mechanics and photonics at the same time. If we adopt conventions in photonics, the Hamiltonian for the TE mode propagating light becomes

$$H = -\frac{1}{k_0^2}\frac{d^2}{dx^2} - \varepsilon(x), \tag{2.3}$$

where $\varepsilon(x)$ is the profile of the relative permittivity and k_0 is the free space wavenumber. The mth eigenvalue γ_m for the eigenvalue equation $H\psi_m = \gamma_m \psi_m$ then represents the mth effective modal index n_m with $\gamma_m = -n_m^2$.

While the SUSY technique can be applied to non-Hermitian Hamiltonians with complex-valued potentials (Miri et al. 2013a, b) such as optical structures with gain and loss media (Feng et al. 2017; El-Ganainy et al. 2018), in this book, we restrict our discussion to the Hermitian case where the permittivity $\varepsilon(x)$ is a real-valued function. The factorization of the Hamiltonian $H = BA + \gamma_d$ is then simplified into $H = A^\dagger A + \gamma_d$ (Miri et al. 2013a, b; Yu et al. 2015a, b, 2017, 2018; Miri et al. 2014). Because the Schrödinger-like Hamiltonian is the function of the second derivative, the intertwining operator A has to be the function of the first derivative, as $A = (1/k_0)\partial_x + V(x)$. The scalar potential $V(x)$ of the intertwining operator A is called the superpotential. The form of the operator A results in the Riccati equation

$$\partial_x V = k_0 V^2 + k_0[\varepsilon(x) + \gamma_d] \tag{2.4}$$

for the superpotential $V(x)$. The particular solution of the Riccati equation for non-diverging $V(x)$ is well-known (Zaitsev 2002), as

$$V(x) = -\frac{1}{k_0}\frac{\partial_x \psi_d}{\psi_d}, \tag{2.5}$$

where ψ_d is defined as the nodeless eigenmode of $H\psi_d = \gamma_d\psi_d$ with $\gamma_d \le \gamma_0$, for the "bound" ground state of $H\psi_0 = \gamma_0\psi_0$.

In SUSY quantum mechanics and SUSY photonics based on the Schrödinger-like Hamiltonian, the condition of $\gamma_d = \gamma_0$ and $\psi_d = \psi_0$ provides the annihilation condition of $A\psi_0 = O$, the *ground state annihilation* in the "unbroken" SUSY Hamiltonian for the bosonic (ψ_m) and fermionic ($A\psi_m$) relationship (Cooper et al. 2001). Therefore, the unbroken SUSY transformation with $A = (1/k_0)\partial_x + V(x)$ and $V(x) = -(1/k_0)\partial_x\psi_0(x)/\psi_0(x)$ does not allow realization of the complete isospectrality between different shapes of potentials; the *quasi*-isospectrality is achieved only for the excited states of the original potential ($m = 1, 2, \ldots$) except the ground state ($m = 0$). In contrast, the broken SUSY (Andrianov et al. 1984a, b) with the condition of $\gamma_d < \gamma_0$ does not lead to the ground state annihilation, achieving complete isospectrality. We focus on the unbroken SUSY for discussion.

Unbroken SUSY transformation Using the intertwining operator $A = (1/k_0)\partial_x + V(x)$, we can achieve the SUSY partner Hamiltonian $H_s = H_{iso} = AA^\dagger + \gamma_d$ with the transformation of eigenmodes $A\psi_m$, which derives the intertwining relationship of $AH\psi_m = (AA^\dagger A + \gamma_d A)\psi_m = (AA^\dagger + \gamma_d)A\psi_m = H_s(A\psi_m) = \gamma(A\psi_m)$ for the isospectrality. When we consider the unbroken SUSY with $\gamma_d = \gamma_0$ and $\psi_d = \psi_0$, the SUSY partner Hamiltonian $H_s = AA^\dagger + \gamma_0$ becomes

$$H_s = -\frac{1}{k_0^2}\frac{d^2}{dx^2} - \left[\varepsilon(x) + \frac{2}{k_0^2}\frac{d}{dx}\left(\frac{\partial_x\psi_0}{\psi_0}\right)\right], \tag{2.6}$$

which leads to the SUSY-transformed optical potential (or permittivity profile)

$$\varepsilon_s(x) = \varepsilon(x) + \frac{2}{k_0^2} \frac{d}{dx} \left(\frac{\partial_x \psi_0}{\psi_0} \right), \tag{2.7}$$

and SUSY-transformed eigenmodes $\psi_{s\text{-}m}(x) = A\psi_m(x)$, or

$$\psi_{s-m}(x) = \frac{1}{k_0} \left[\frac{d}{dx} - \left(\frac{\partial_x \psi_0}{\psi_0} \right) \right] \psi_m(x) \tag{2.8}$$

with $\psi_{s\text{-}0}(x) = A\psi_0(x) = O$.

Other sets of isospectral potentials can be obtained from the unbroken SUSY partner ($\gamma_d = \gamma_0$), which are called isospectral families (Cooper et al. 2001; Miri et al. 2014). From the unbroken SUSY Hamiltonian $H_s = AA^\dagger + \gamma_0$ with the SUSY optical potential $\varepsilon_s(x)$, the Riccati equation for the "SUSY" potential $\varepsilon_s(x)$, not for the original potential $\varepsilon(x)$ in Eq. (2.4), has the form $\partial_x V = -k_0 V^2 - k_0(\varepsilon_s + \gamma_0)$. Using the known particular solution of $V(x) = -(1/k_0)\partial_x \psi_0(x)/\psi_0(x)$, the general solution is obtained as

$$V(x) = -\frac{1}{k_0} \frac{\partial_x \psi_0(x)}{\psi_0(x)} + \frac{\psi_0^2(x)}{c + \int_{-\infty}^{x} k_0 \psi_0^2(x)dx}, \tag{2.9}$$

where c is an arbitrary constant ($c \to \pm\infty$ for the original unbroken SUSY). From the superpotential $V(x)$ in Eq. (2.9), other isospectral potentials of $\varepsilon(x)$ are obtained as $\varepsilon_f(x) = -V^2 + (1/k_0)\partial_x V - \gamma_0$.

SUSY Photonics From the above discussion, we summarize **two key properties of the unbroken SUSY transformation**, which can be compared with the properties of the isospectral transformations in Sect. 2.1 (**P- for property**):

P-I. Conserved eigenspectrum γ_m ($m = 1, 2, \ldots$), except for the eigenvalue γ_0 of the original ground state $\psi_0(x)$, and
P-II. Transformed eigenmode ($\psi_m \to A\psi_m$) and **ground state annihilation** ($A\psi_0 = O$) with the transformed potential (or Hamiltonian $AH = H_sA$), governed by the modal profile of the original ground state $\psi_0(x)$.

Thus, the key feature of the unbroken SUSY transformation is the achievement of *quasi*-isospectrality with the ground-state annihilation. This result is based on the modulation of the optical potential $\varepsilon(x)$ according to the spatial profile of the ground state $\psi_0(x)$, as shown in Eq. (2.7). The selective control of the eigenspectrum, by removing a single ground state while preserving the phase matching condition in the other states, has provided a powerful tool in the design of photonic devices, especially for the handling of multimode photonic structures, which is called the field of SUSY photonics (Miri et al. 2013a, b; Heinrich et al. 2014a, b; Yu et al. 2015a, b, 2017, 2018; Hokmabadi et al. 2019; Longhi 2015a, b; Barkhofen et al. 2018; Miri et al. 2013a, b, 2014; Midya et al. 2018; Smirnova et al. 2019; Queraltó

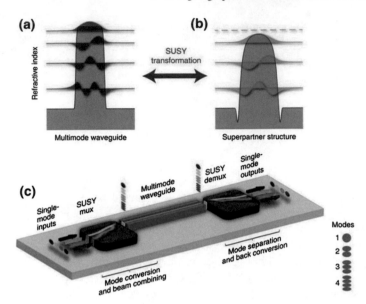

Fig. 2.4 Concept of SUSY mode converters for mode division multiplexing (MDM). A multimode optical waveguide and its SUSY partner are shown in **a** and **b**, allowing the photonic MDM device in **c** from a hierarchical sequence of multiple superpartner structures Figure adapted from Heinrich et al. (2014a, b) under a CC BY license (http://creativecommons.org/licenses/by/4.0/)

et al. 2018; Teimourpour et al. 2016; Principe et al. 2015; Chumakov and Wolf 1994; Laba and Tkachuk 2014; Heinrich et al. 2014a, b; Walasik et al. 2018; Macho et al. 2018; El-Ganainy et al. 2015; Zuniga-Segundo et al. 2014; Longhi 2014).

The first application of SUSY in photonics was shown in the interpretation of optical waveguides in terms of Helmholtz optics (Chumakov and Wolf 1994). The concept of SUSY for photonics was revisited by Miri et al. 2013a, b to design multimode devices using the ground state annihilation (Miri et al. 2013a, b). Including the experimental verification of SUSY mode converters (Fig. 2.4) (Heinrich et al. 2014a, b), the isospectral nature of the SUSY transformation has been applied to various interesting phenomena and applications including non-parity-time-symmetric complex potentials with real spectra (Miri et al. 2013a, b), disordered potentials with perfect bandgaps (Yu et al. 2015a, b), random wave switching (Yu et al. 2017, 2018), Talbot imaging (Longhi 2014), and recently, SUSY laser arrays (Hokmabadi et al. 2019).

From the concept and properties of the unbroken SUSY transformation, we introduce the applications of SUSY isospectrality in disordered photonics, including disordered photonic structures for perfect photonic bandgaps (Sect. 2.3) and controlling random waves in SUSY photonic structures for logic devices (Sect. 2.4).

2.3   Application I: Supersymmetric Disorder for Photonic Bandgaps

Bloch's theorem in 1928 was a major milestone in solid-state physics (Kittel et al. 1976; Marder et al. 2010). By allowing the description of waves in periodic structures, *a.k.a.* Bloch waves, Bloch's theorem established the principle of "bandgaps" in solids with rigorous mathematical descriptions that were used to roughly classify solids into metals, semiconductors, and insulators according to their bandgap properties. The utility of bandgap phenomena is not restricted to electrical transport; this concept has also been applied to other physical waves, such as light in photonic crystals and sound in phononic crystals. Before the discovery of amorphous media in Clark (1967), it was strongly believed that bandgaps could be obtained only in crystals that exhibit "periodicity" and "long-range correlations", which were regarded as prerequisites for Bloch's theorem.

The belief in the requirements of "long-range correlations" and "periodicity" for Bloch waves persisted for almost 40 years before being disproven by the ground-breaking discoveries of amorphous media (Clark 1967) and quasicrystals (Levine and Steinhardt 1984). Because Bloch's theorem cannot be applied to aperiodic structures, diverse approaches have been suggested to interpret the counterintuitive observations of "Bloch-like waves in disorder", e.g., wide and perfect bandgaps in disordered potentials. Most of these approaches have used "iterative" or "statistical" searches for random structures with bandgaps, lacking the deterministic creation of bandgap media based on fundamental wave equations.

The goal of this section is the determination of the isospectral relations between order (crystals or quasicrystals) and a certain type of designed disorder to realize wide and perfect bandgaps in disordered potentials (Yu et al. 2015a, b). Before employing the SUSY transformation in Sect. 2.2, we introduce the mathematical tool for measuring the long-range correlation of disordered potentials, the *Hurst exponent* (Yu et al. 2015a, b; Hurst 1951; Roche et al. 2003; King et al. 2017), defined by the rescaled range (R/S) analysis (Mandelbrot and Wallis 1969).

Hurst Exponent Consider the spatial profile function of the refractive index $n(x)$ of an optical potential. To quantify the long-range correlation of the given potential landscape with a "single" statistical parameter, we first discretize the continuous function $n(x)$. Then, we obtain the discretized refractive index n_p ($p = 1, 2, \ldots,$ N) with the N-element discretization at $x_p = x_{\text{left}} + (p - 1)\Delta$, where x_{left} is the left boundary of the potential, which is of length $L = (N - 1)\Delta$. From n_p, partial sequences X_q for different length scales d can be defined ($2 \leq d \leq N$ and $1 \leq q \leq d$); for example, for the discretized refractive index sequence $n_p = \{n_1, n_2, \ldots, n_N\}$, we can define the partial sequence $\{X_1, X_2, X_3, X_4\} = \{n_9, n_{10}, n_{11}, n_{12}\}$. For the mean-adjusted sequence $Y_q = X_q - m$, where m is the mean of X_q, we achieve the cumulative deviate series Z_r:

$$Z_r = \sum_{q=1}^{r} Y_q. \tag{2.10}$$

The range of cumulative deviation is defined as $R(d) = max(Z_1, Z_2, ..., Z_d) - min(Z_1, Z_2, ..., Z_d)$, which represents the range of the first d cumulative deviations from the mean. Using the standard deviation $S(d)$ of Y_q, we apply the power law to the rescaled range $R(d)/S(d)$ as follows

$$E\left[\frac{R(d)}{S(d)}\right] = c_0 d^H. \tag{2.11}$$

This relationship yields $log(E[R(d)/S(d)]) = H \cdot log(d) + c_1$, where E is the expectation value and c_0 and c_1 are constants. The parameter H is then obtained through linear polynomial fitting, which is called the Hurst exponent. The Hurst exponent H quantifies the long-range order (Hurst 1951; Roche et al. 2003) or long-term memory of the given sequence: $H = 0.5$ for Brownian motion, $0 \leq H < 0.5$ for long-term negative correlations with switching behaviours and $0.5 < H \leq 1$ for long-term positive correlations such that the sign of the signal is persistent or has less roughness.

Structural Correlations and Spectral Information of Light Using the statistical analysis based on the Hurst exponent H, we can examine the relationship between the eigenspectra and the structural correlations of optical potentials for three types of potentials, crystals, quasicrystals, and random potentials, which are generated by adjusting the refractive index profile. Figure 2.5a represents an example of 1-dimensional binary Fibonacci quasicrystals (the 6th generation with an inflation number, or sequence length, of $N = 8$, substituting A $\rightarrow$ B and B $\rightarrow$ BA for each generation using A as the seed). Each element is defined by the gap between the high-index regions: A (or B) for a wider (or narrower) gap. The crystal and the random potential are then generated using the same definition of elements, while the crystal has an alternating sequence (BABABA...), and the random potential has equal probabilities of A and B for each element that corresponds to a Bernoulli random sequence with probability $p = 0.5$. To quantify the correlation, the Hurst exponent H is introduced (Fig. 2.5b). As N increases, both the crystal and the quasicrystal have H values that approach 0, exhibiting "ballistic behaviour" with strong negative correlations. In contrast, the Bernoulli random potential has $H \sim 0.4$, being close to ideal Brownian motion with $H = 0.5$.

Figure 2.6a–c illustrate the stationary eigenmodes for each potential. Consistent with previous studies (Kittel et al. 1976; Weaire and Thorpe 1971; Edagawa et al. 2008; Rechtsman et al. 2011; Joannopoulos et al. 2011; Vardeny et al. 2013; Wiersma 2013), wide and perfect bandgaps are allowed only for the ordered potentials of the crystal (Fig. 2.6a, d) and the quasicrystal (Fig. 2.6b, e), and these bandgap spectra

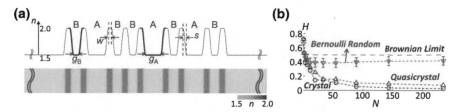

Fig. 2.5 Relationship between eigenmodes and potential correlations. **a**, Definitions of elements, illustrated for an example of a 1D Fibonacci quasicrystal ($N = 8$). $g_A = 600$ nm, $g_B = 200$ nm, $w = 120$ nm, $s = 140$ nm, and the wavelength is $\lambda_0 = 1500$ nm. The distance between high refractive index regimes g_A and g_B defines an element for a quasicrystal. **b** Hurst exponent H for each potential as a function of the sequence length N. H quantifies the strength of disorder. The sequence lengths N are selected to be equal to those of Fibonacci quasicrystals. The H of the Bernoulli random potential is plotted with standard deviation error bars for 200 statistical ensembles. The black dashed line represents the Hurst exponent of ideal Brownian motion ($H = 0.5$) Figure and caption adapted from Yu et al. (2015a, b) under a CC BY license (http://creativecommons.org/licenses/by/4.0/)

become more apparent with increasing N (Figs. 2.6d, e). By contrast, no bandgap is observed for the Bernoulli random potential, which lacks *any* correlations (Fig. 2.6c), especially for larger N (Fig. 2.6f). This lack of correlation originates from the broken coherence of this case, which hinders the destructive interference that is necessary for the formation of bandgaps. Many eigenmodes are localized within this random potential, exhibiting a phenomenon that is widely known as Anderson localization (Anderson 1958).

In light of the results in Fig. 2.6, we now consider the following question: "*Is it possible to design Brownian potentials with H~0.5 that possess wide and perfect bandgaps?*". To answer this question in the affirmative, we exploit the SUSY transformation to achieve the family of quasi-isospectral potentials.

SUSY-induced Disorder Fig. 2.7 illustrates an example of serial SUSY transformations applied to the 1D Fibonacci quasicrystal potential defined in Figs. 2.5 and 2.6, where the small value of $N = 5$ is selected for clarity of presentation. For each SUSY transformation, all eigenmodes of each previous potential, except for the ground state (or highest n_{eff} state), are preserved in the transformed spatial profiles, while the shape of the designed potential becomes "disordered" through "deterministic" SUSY transformations. Therefore, the SUSY transformation, which is defined by the ground-state-based modulation of the potential, leads to disordered optical potentials with a bandgap identical to that of the original crystal, in contrast to "random" potentials without bandgaps.

Regardless of whether deterministic (Yu et al. 2015a, b) or statistical (Florescu et al. 2009; Torquato 2016; Man et al. 2013; Hejna et al. 2013; Florescu et al. 2013; Torquato and Stillinger 2003; DiStasio et al. 2018), the presence of a certain type of order is essential for achieving destructive interferences and the resulting bandgaps. The "*random-like* shapes" of potentials (not actually *random*, Fig. 2.7) "deterministically" derived by applying SUSY transformations to ordered potentials thus offer the possibility of combining bandgap phenomena and disordered potentials. We

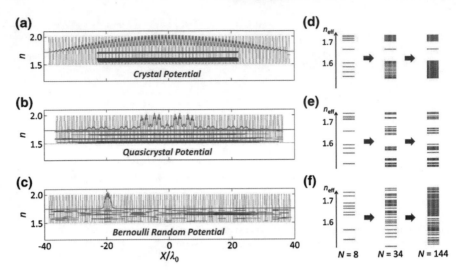

Fig. 2.6 a–c, Eigenstates of each potential. The blue curve represents the ground state of each potential, and the coloured lines represent the spectral (n_{eff}) distributions of the eigenmodes. **d–f** Evolutions of the band structures for different sequence lengths N: **a, d** for crystals, **b, e** for quasicrystals, and **c, f** for Bernoulli random potentials. Note that the eigenstate inside the gap in **a, d** is a surface state for an even N (or an odd number of high-index regions) from the finite sizes of the potentials Figure and caption adapted from Yu et al. (2015a, b) under a CC BY license (http://creativecommons.org/licenses/by/4.0/)

consider a larger-N regime in which the wave phenomena and spectral information are clearly distinguished between ordered and disordered potentials. Figures 2.8a, b show the results achieved after the 10th unbroken SUSY transformation for the crystal (Figs. 2.8a, c) and the quasicrystal (Figs. 2.8b, d). Although the shapes of the SUSY-transformed potentials and the profiles of the eigenmodes in Fig. 2.8a, b are markedly different from those of the corresponding original potentials in Fig. 2.6, the eigenspectrum of each potential is preserved, save for the annihilation of the 10 lowest eigenmodes: successive *ground state annihilation*. Figures 2.8c, d depict the variation in the effective modal index that occurs during the SUSY transformations, showing that the eigenspectrum of each potential is maintained from the original to the 20th SUSY transformation, save for a shift in the modal number, and therefore, the bandgaps in the remainder of the spectrum are maintained (~125 states after the 20th SUSY transformation, following the loss of the 20 annihilated states). Consequently, bandgaps identical to those of the original potentials are achieved in SUSY-transformed potentials despite their disordered shapes. These disordered potentials (Fig. 2.3a, b) can be classified as neither crystals nor quasicrystals.

Statistical Analysis of SUSY Disorder Figures 2.9a–h illustrate the evolutions of the shapes of the SUSY-transformed crystal and quasicrystal potentials, demonstrating an increase in disorder (or structural modulation) for both potentials. Note that the ground-state profile $\psi_0(x)$ is concentrated near the centre of the potential (Fig. 2.6a, b), representing the direction of the structural modulation.

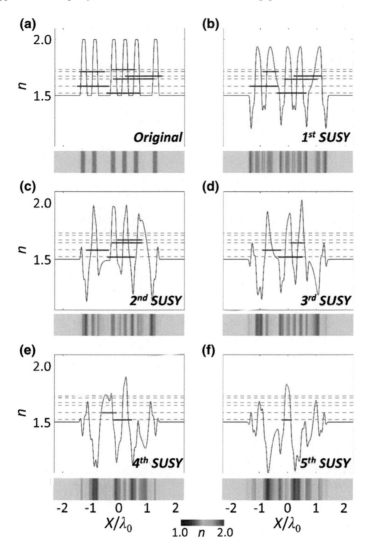

Fig. 2.7 SUSY transformation for quasi-isospectral potentials. A 1D Fibonacci quasicrystal ($N = 5$) is considered as an example. **a** Original potential. **b–f** 1st–5th SUSY-transformed potentials. The orange (or black) dotted lines represent the preserved (or annihilated) eigenmodes Figure and caption adapted from Yu et al. (2015a, b) under a CC BY license (http://creativecommons.org/licenses/by/4.0/)

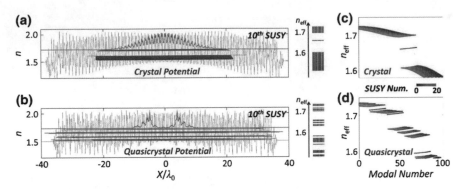

Fig. 2.8 Eigenmodes of SUSY-transformed crystals and quasicrystals. The 10th SUSY-transformed potentials and their eigenmodes are depicted for **a** a crystal potential and **b** a quasicrystal potential. **c** and **d** show the eigenvalues of the SUSY-transformed potentials as a function of the modal numbers of the crystal and quasicrystal potentials, respectively. The 0th SUSY-transformed potential corresponds to the original potential. $N = 144$ Figure and caption adapted from Yu et al. (2015a, b) under a CC BY license (http://creativecommons.org/licenses/by/4.0/)

To quantify the correlation features of SUSY-transformed potentials, we use the Hurst exponent. Figures 2.10a, b show the Hurst exponents for the SUSY-transformed crystal and quasicrystal potentials as functions of the number of SUSY transformations, for different sequence lengths ($N = 34, 59, 85$, and 144). The results show that the Hurst exponents of the crystal and quasicrystal potentials ($H = 0$–0.1) increase and saturate at $H \sim 0.8$ for successive applications of SUSY transformations. At $N = 144$, the negative correlations ($H < 0.5$) of the crystal and quasicrystal potentials ($H = 0$–0.1) become almost uncorrelated, with $H = 0.51$ after the 10th SUSY transformation, the correlations are even weaker than that of the Bernoulli random potential ($H = 0.35$–0.48) and approach the uncorrelated Brownian limit $H = 0.5$. After the 10th SUSY transformation,, the correlation again begins to increase into the positive-correlation ($H \geq 0.5$, with long-lasting shapes), exhibiting a transition between negative and positive correlations. This $\psi_0(x)$-dependent modulation shows the size (or N) dependence; for a potential of large size, $\psi_0(x)$ varies weakly over a wide range, thus decreasing the relative strength of the SUSY-induced modulation. Thereby, the number of SUSY transformations, required for maximum randomness ($H \sim 0.5$) increases with N, $(S_B, N) = (4, 34), (6, 55), (8, 89)$, and $(10, 144)$, where S_B is the required transformations for $H \sim 0.5$. Eventually, the SUSY transformation, to periodic potentials of *infinite* size $n(x) = n(x + \Lambda)$ preserves the periodicity because the SUSY transformation, with the Bloch ground state $\psi_0(x) = \psi_0(x + \Lambda)$ repeatedly results in periodic potentials $n_s(x) = n_s(x + \Lambda)$.

Thus, the application of SUSY transformations, to ordered potentials allows remarkable control of the extent of disorder while preserving bandgaps. A family of potentials with "Bloch-like eigenvalues", because the members of this family have identical bandgaps but tunable disorder, can be constructed through the successive application of SUSY transformations, to each ordered potential, with a range of dis-

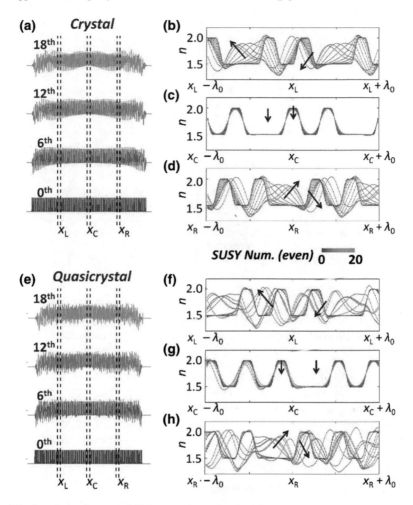

Fig. 2.9 Correlation features of SUSY-transformed potentials. The evolutions of the potential profiles following the successive application of SUSY transformations, (0th, 6th, 12th, and 18th) for **a** a crystal and **e** a quasicrystal. **b–d** and **f–h** present magnified views (at x_L, x_C, and x_R) of the landscapes of the potentials for even numbers of SUSY transformations, (overlapping up to the 20th transformation; the blue arrows indicate the direction of potential modulation). $N = 144$ in **a–h** Figure and caption adapted from Yu et al. (2015a, b) under a CC BY license (http://creativecommons.org/licenses/by/4.0/)

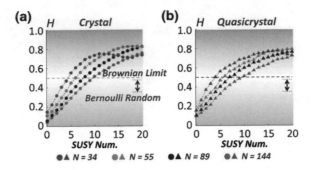

Fig. 2.10 Hurst exponents H as functions of the number of SUSY transformations, are shown for **a** crystals and **b** quasicrystal potentials, with different sequence lengths ($N = 34, 59, 85$, and 144). The red (or blue) region represents the regime of positive (or negative) correlation, whereas the white region corresponds to the uncorrelated Brownian limit. The arrow indicates the regime of the Bernoulli random potential, which represents the disorder strength of binary random structures Figure and caption adapted from Yu et al. (2015a, b) under a CC BY license (http://creativecommons. org/licenses/by/4.0/)

order spanning almost the entire regime of Hurst exponents indicating negative and positive correlations ($0 \leq H \leq 0.8$), including the extremely uncorrelated Brownian limit of $H \sim 0.5$.

Modal Localization From the SUSY transformation, by Eq. (2.8), we expect that the distribution of the ground state $\psi_0(x)$ with respect to the original state ψ primarily affects the effective width of the transformed eigenmode $\psi_{s-m}(x) = A\psi_m(x)$. Figure 2.11 shows examples of SUSY-transformed eigenmodes (from the 0th to the 3rd modes) in crystals and quasicrystals. For a crystal with eigenmodes that have highly overlapping intensity profiles, the bound profile of $\psi_0(x)$ decreases the spatial bandwidth of each eigenmode with the contribution of $(\partial_x \psi_0/k_0\psi_0)\psi$. However, because the eigenmodes in the quasicrystal are already spatially separated, in contrast to those in the crystal, the spatial modification by $\psi_0(x)$ occurs in a much more complex manner.

To quantify the localization of eigenmodes in SUSY-transformed potentials, we introduce the definition of the effective width w_{eff} based on the inverse participation ratio (Schwartz et al. 2007) as

$$w_{\text{eff}} = \frac{\left[\int I(x)dx\right]^2}{\int I^2(x)dx},\qquad(2.12)$$

where $I(x)$ is the intensity of each eigenmode. Figure 2.12 shows the change in the effective width for the 30 lowest eigenvalue modes by serial SUSY transformations. As shown, the effective width gradually decreases in the crystal potentials, whereas no tendency is observed in the quasicrystal potentials. However, for both crystals and quasicrystals, localized eigenmodes were found from serial SUSY transformations.

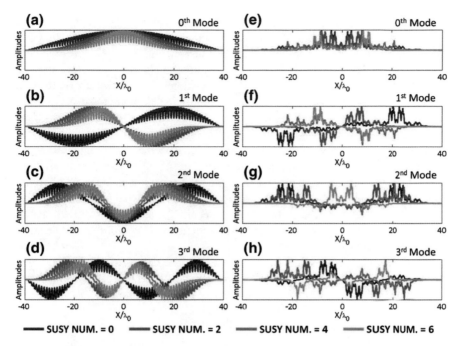

Fig. 2.11 Spatial variations of eigenmodes from the successive applications of SUSY transformations, (black: 0, green: 2, blue: 4, and orange: 6 SUSY transformations,): crystals (**a–d**) and quasicrystals (**e–f**) for different mode numbers (**a, e**: 0th, **b, f**: 1st, **c, g**: 2nd, and **d, h**,: 3rd mode) Figure and caption adapted from Yu et al. (2015a, b) under a CC BY license (http://creativecommons.org/licenses/by/4.0/)

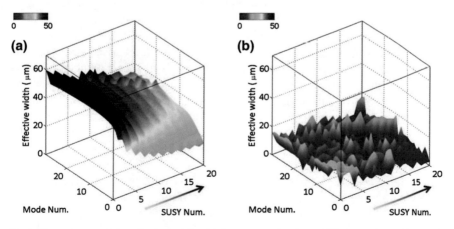

Fig. 2.12 Variation of the effective widths of the eigenmodes from SUSY transformations: **a** in the crystal, and **b** in the quasicrystal. The number of SUSY transformations, changes from 0 to 20. $N = 144$, and the initial shapes of potentials are same as those used in the main manuscript Figure and caption adapted from (Yu et al. 2015a, b) under a CC BY license (http://creativecommons.org/licenses/by/4.0/)

Extension to 2-dimensional Structures Could we achieve higher-dimensional disordered potentials using the SUSY transformation? It is not so simple. In stark contrast to the case of 1D potentials, which exclusively have a 1:1 correspondence between their shape and ground state, it is more challenging to achieve isospectrality in multidimensional problems. Although studies have shown the vector-form SUSY decomposition of multidimensional Hamiltonians (Andrianov et al. 1984a, b; Cannata et al. 2003), such an approach, which is analogous to the Moutard transformation (Gutshabash 2013), cannot guarantee isospectrality. This approach only generates a pair of scalar Hamiltonians with eigenspectra that, in general, do not overlap but together compose the eigenspectrum of the other vector-form Hamiltonian. While such an approach provides insight into the extension of SUSY photonics to vector fields, we introduce an example from an alternative and much simpler route (Kuru et al. 2001; Demircioğlu et al. 2002; Yu et al. 2015a, b; Longhi 2015), starting from the intertwining relation $AH = H_s A$ to implement a class of multidimensional isospectral potentials.

The basic principle is based on the separation of variables to reduce the dimensionality of the governing equation. The procedure of the 1D SUSY transformation, shown in Sect. 2.2 can be independently applied to a 2D potential for each x- and y-axis when the original potential satisfies the condition $V_o(x, y) = V_{ox}(x) + V_{oy}(y)$. For such types of potentials, the series of 2D SUSY transformations are allowed because the form of $V_o(x, y) = V_{ox}(x) + V_{oy}(y)$ is preserved during the transformation. Consequently, such as for the 1D problems, we can derive a family of 2D quasi-isospectral potentials from the original potential.

Figure 2.13 shows an example of SUSY transformations, in 2D potentials, maintaining the original bandgaps. Both the x- and y-axis cross-sections of the 2D original potential $V_o(x, y) = V_{ox}(x) + V_{oy}(y)$ have profiles of $N = 8$ binary sequences. We then apply SUSY transformations, separately to the x- and y-axes, which leads to the highly anisotropic shape of the potential as shown in Fig. 2.13b (the 5th x-axis SUSY transformed potential) and Fig. 2.13c (the 5th y-axis SUSY transformed potential). This anisotropy can be controlled by independently changing the numbers of SUSY transformations, for the x- and y-axes. The isotropic application of SUSY transformations, then recovers the isotropic potential shape (Fig. 2.13d). Regardless of the number of SUSY transformations, and their anisotropic implementations, the region of bandgaps of the original potential is always preserved (Fig. 2.13e) according to the *quasi*-isospectrality of the unbroken SUSY transformation. Interestingly, annihilation by 2D SUSY transformation can occur not only in the ground state but also in all of the excited states sharing a common 1D ground-state profile. Therefore, the width of the bandgap can be slightly changed owing to the annihilation of certain excited states near the bandgap.

To investigate the correlation features of 2D SUSY-transformed potentials, we quantify the angle-dependent degree of the correlation, again by using the Hurst exponent H. Figures 2.13f, g represent the angle-dependent variation of the Hurst exponent for the anisotropic (the 5th x-axis SUSY-transformed potential, Fig. 2.13b) and isotropic (the 5th x- and y-axis SUSY transformed potential, Fig. 2.13d) disordered potentials. Compared with the original potential (grey symbols in Fig. 2.13f,

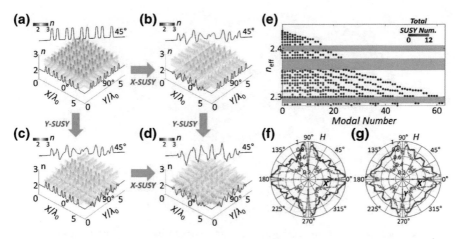

Fig. 2.13 2D SUSY-transformed potentials with bandgaps maintained. The evolutions of the potential profiles following the application of SUSY transformations, to the x- and y-axes are shown: **a** original and **b** x-axis SUSY transformed (5th). **c**, y-axis SUSY transformed (5th) and **d,** x- and y-symmetric SUSY-transformed (x-axis: 5th, y-axis: 5th) potentials. The spatial profiles of the potentials for $0°$ (x-axis), $45°$, and $90°$ (y-axis) are also overlaid in **a–d**. **e** shows the eigenvalues of the SUSY-transformed potentials as a function of the modal numbers. The grey regions denote bandgaps. The total SUSY number is the sum of the numbers of SUSY transformations, for x- and y-axes (i.e., 5 for both **b** and **c** and 10 for **d**). **f** and **g** are the Hurst exponents for different directions of the 2D potentials that are **f** highly anisotropic (x-axis: 5th, y-axis: 0th) and **g** quasi-isotropic (x-axis: 5th, y-axis: 5th) SUSY transformations. The grey symbols in **f** and **g** are the Hurst exponents of the original potential Figure and caption adapted from (Yu et al. 2015a, b) under a CC BY license (http://creativecommons.org/licenses/by/4.0/)

g, for Fig. 2.13a), H increases along the axis with the SUSY transformations, (x-axis in Fig. 2.13f and x- and y-axes in Fig. 2.13g). The potential is disordered at all angles, especially in the diagonal directions ($\pm 45°$), owing to the projection of the SUSY-induced disordered potential shapes ($45°$ profiles in Fig. 2.13b–d).

What is SUSY Disorder? To summarize, by employing SUSY transformations, we find a new path towards the deterministic creation of disordered potentials with "crystal-like" spectral information and tunable structural correlations, extending the frontier of disorder for perfect bandgaps. Despite their weak spatial correlations and disordered shapes, SUSY-transformed potentials retain the deterministic "ground-state-dependent order", which is the origin of bandgaps. This result is deterministically derived based on a Schrodinger-like wave equation, without any trial-and-error techniques or statistical approaches. The most critical impact of the SUSY disorder is the achievement of (i) the spectral property of crystals (or bandgaps) and (ii) the modal property of random potentials (or modal localization) at the same time as shown in Fig. 2.14, in sharp contrast to the regime of the monotonic transition between order and randomness.

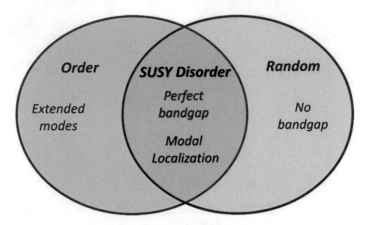

Fig. 2.14 Concept of SUSY disorder: anomalous intermediate regime between order and randomness, simultaneously allowing perfect bandgaps and modal localization

2.4 Application II: Supersymmetric Switching of Random Waves

The remarkable success of modern electronics is due in large part to the stability of charge-based signal processing. For example, a transistor, the most fundamental element in digital electronics, operates with current flow that is controlled by an electromagnetic field exerted on electrons (Streetman 2000), regardless of the "phase information" of the electric wave functions. Therefore, the electronic switching is insensitive to the specific wave function profiles, allowing a high degree of freedom for information capacity.

On the other hand, controlling light for signal processing is achieved only through light-matter interactions. Because light-matter interactions are defined by light fields (both amplitude and phase functions) interacting with optical potentials, the modulation of light strongly depends on specific modal profiles, which hinders consistent dynamics. This dependence is one of the key bottlenecks in achieving "active" multichannel photonic devices. Despite the continued success of passive multimode devices for high information capacity (Heinrich et al. 2014a, b; Mrejen et al. 2015; Wright et al. 2015; Yu et al. 2008, 2013), most state-of-the-art optical switching has employed single mode operation (Li et al. 2016; Ganesh and Gonella 2015; Nozaki et al. 2010; Yu et al. 2011, 2015; Piao et al. 2011, 2012, 2015) with multiplexers/demultiplexers (Fig. 2.15a), and studies on multimode switching devices are very rare (Yu et al. 2012). Thus, the realization of collective switching of multimodes is an urgent issue for full access to the high bandwidth of light; could we achieve switching of "random light" regardless of the random amplitude and phase functions of this type of light (Fig. 2.15b)? In the context of disordered photonics, this subject is concerned with the control of "disorder in the light itself", not in the optical potentials.

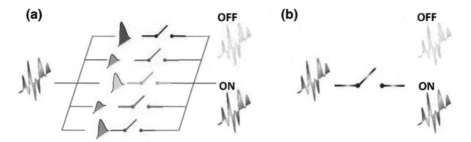

Fig. 2.15 a Conventional switching with multiplexer/demultiplexers and multiple single mode switches. **b** A switching building block for random waves. Reprinted figure with permission from Yu et al. (2017, 2018). Copyright 2017 by the APS

Design Criteria The first goal is the construction of a building block for collective multimode switching (Fig. 2.15b) that is transparent to the spatial field profiles of incident light. First, the "simultaneous" switching of all the participating modes is necessary. This condition requires the eigenspectra in the input and output (I/O) potentials to be *equally spaced* and *identical*. While the condition of *equally spaced* spectra requires harmonic or Wannier-Stark (WS) ladder (Sapienza et al. 2003) potentials, the *identical* spectra, or *isospectrality* in Sect. 2.1, can be implemented through the SUSY transformation, in Sect. 2.2. Second, to achieve perfect isolation in the "off" state, all the eigenmodes in the input and output potentials should have zero modal overlap. To summarize, **the design criteria** include (**C- for criterion**):

C-I. Isospectral pair of I/O ports composed of harmonic or WS ladder potentials and
C-II. Orthogonal condition between all eigenmodes inside I/O ports.

To satisfy these two conditions, we introduce the concept of "parity-reversed contact" (Yu et al. 2011) of input and output potentials, with spectrally matched but parity-reversed eigenmodes (Fig. 2.16a, b), for the isospectrality with equally spaced spectra and perfectly decoupled eigenmodes.

Parity-Reversed SUSY Potentials In the application of Sect. 2.3, the realization of disordered potentials with perfect bandgaps, we focused on the isospectrality of the unbroken SUSY transformation. However, as shown in Eq. (2.8), the application of the SUSY transformation, also results in the transformation of the spatial modal profile. In this section, we pay more attention to the use of the modal property of the SUSY transformation, to construct the building block for multimodal "parity conversion". First, we consider the parity operator P with $Pf(x) = f(-x)$ and a parity-symmetric potential $\varepsilon(x) = \varepsilon(-x)$. Because the Hamiltonian H satisfies the commutation $[H, P] = HP - PH = O$, all the eigenmodes have definite parity as $\psi(x) = \pm \psi(-x)$. Thus, the ground state ψ_0, which is nodeless, has even parity with $\varepsilon(x) = \varepsilon(-x)$. Then, the unbroken SUSY operator $A = (1/k_0)[\partial_x - \partial_x \psi_0(x)/\psi_0(x)]$ composed of $\partial_{-x} = -\partial_x$ and even-function ground state $\psi_0(x) = \psi_0(-x)$ derive the "parity reversal" of all the eigenmodes for an original parity-symmetric potential

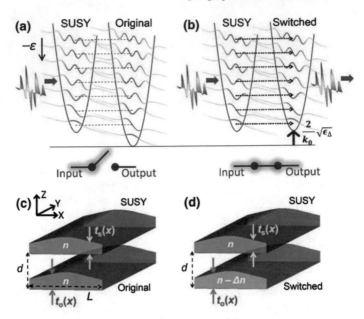

Fig. 2.16 Building block for binary switching of random waves. **a** Off-state. **b** On-state. Eigenmodes are plotted on the potentials (red: even, blue: odd) in **a**, **b**. The corresponding waveguides are shown in **c** (off-state) and **d** (on-state). Reprinted figure with permission from Yu et al. (2017, 2018). Copyright 2017 by the American Physical Society (APS)

$\varepsilon(x) = \varepsilon(-x)$. The unbroken SUSY transformation, allows the "collective parity conversion" with the isospectrality for *any* parity-symmetric potential $\varepsilon(x) = \varepsilon(-x)$.

Therefore, among two types of potentials with *equally spaced* spectra, harmonic or Wannier-Stark ladder, we select the harmonic potential, which is initially parity-symmetric. The pairing of a harmonic potential and its SUSY partner is designed (Fig. 2.16a, b) for the I/O ports of the binary switch, which successfully satisfy the design criteria of the collective multimode switching. To summarize, the building block for the binary switching of random waves is constructed by satisfying the two design criteria above (C-I & C-II) with **the following properties of the unbroken SUSY transformation (P- for property)**:

P-I. Isospectral and equally spaced eigenspectra in the SUSY harmonic pair and **P-II. Parity reversal with the SUSY transformation** of the harmonic potential.

As an example of parity-reversed SUSY potentials, we design TE-mode coupled waveguides (Fig. 2.16c); the effective permittivities of the original and SUSY partner harmonic waveguide are $\varepsilon_o(x) = 10.38 - 7.48 \times 10^{-4} \cdot (k_0 x)^2$ and $\varepsilon_s(x) = 10.32 - 7.48 \times 10^{-4} \cdot (k_0 x)^2$, respectively. These effective permittivity functions are realized with the spatially-varying thickness of the dielectric waveguides. The change of the refractive index (Fig. 2.16d) results in the collective transition of the eigenmodes with the parity reversal.

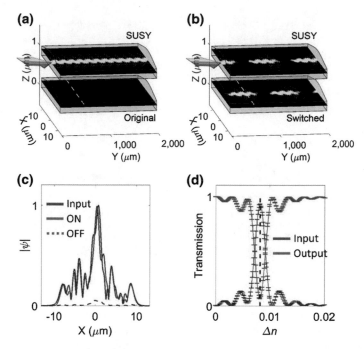

Fig. 2.17 Binary random wave switching in the SUSY harmonic pair: **a**, off and **b**, on states. **c**, The reconstruction of the input wavefront in the output (dotted lines in **a**, **b**). **d** The power distribution in the input and output waveguides as a function of the modulation Δn. Error bars denote the standard deviation for the ensembles of 400 random input waves. Reprinted figure with permission from Yu et al. (2017, 2018). Copyright 2017 by the APS

Binary Switching of Random Waves Fig. 2.17 shows the operation of the mode-independent binary switching of random waves. To examine random combinations of eigenmodes in terms of amplitude and phase, the input wave is defined as $\xi(x, z) = \sum u[0, 1] exp(i \cdot u[0, 2\pi]) \psi_m(x, z)$, where $u[a, b]$ is a random number between a and b following the uniform distribution. Even with the arbitrary input $\xi(x, z)$, the entire eigenmodes of the guided part $\varphi(x, z) = \sum c_m \cdot \psi_m(x, z)$, where $c_m = \int\int \psi_m^*(x, z) \cdot \xi(x, z) dxdz$, are collectively switched by applying a moderate value of index modulation $\Delta n = 8 \times 10^{-3} \sim \Delta n_{\text{eff}}$ independent from the detailed phase or amplitude distributions of $\xi(x, z)$. Moreover, due to the equally-spaced eigenspectra of the harmonic potentials, we achieve the reconstruction of the input wavefronts at the output port (Gordon 2004), first achieving perfect transfer of a "complex field" between separated optical elements without *any* corruption. Additionally, with perfect isolation conditions due to the parity reversal, the high performance of the switching is demonstrated in the modulation depth (26 dB), power transfer (93%), and stability (1.1% error).

Fuzzy Logic Function for Random Waves Extending the application of SUSY transformations, to harmonic potentials, we now introduce another solvable potential of the SUSY Pöschl-Teller (Cooper et al. 2001) pair to achieve multilevel switching.

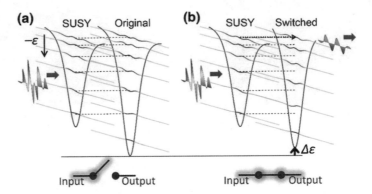

Fig. 2.18 Multilevel switching in the SUSY Pöschl-Teller pair. **a** The off state from the decoupling between the Pöschl-Teller potential and its SUSY partner. **b** One of the many-valued 'on' states from the coupling (black dotted arrow) through the modulated output (red arrow). Reprinted figure with permission from Yu et al. (2017, 2018). Copyright 2017 by the APS

Figure 2.18a, b shows the concept of multilevel switching with the SUSY Pöschl-Teller pair waveguide of $sech^2$ potentials. In the off-state, complete isolation from the isospectral parity reversal is achieved (Fig. 2.18a), such as in the harmonic case. When the modulation increases, the linearly-varying eigenspectrum separation of the Pöschl-Teller potential enables the "sequential" matching of the parity and wavevector between I/O ports (Fig. 2.18b). Such a chain of eigenmodal "transparency" allows the many-valued "true", while the "isolation" represents the "false" (Ross et al. 2009).

We construct the Pöschl-Teller potential by controlling the thickness of dielectric waveguides. The Pöschl-Teller potential with spatially-varying thickness has an effective permittivity distribution of $\varepsilon(x) = \varepsilon_b + \varepsilon_\Delta \cdot sech^2(\alpha x)$, which allows the eigenvalues of $\gamma_m = -\varepsilon_b - (\alpha/k_0)^2 \cdot (p - m)^2$ $(m = 0, 1, ..., floor(p))$, where $p = -(1/2) + [(1/4) + (k_0/\alpha)^2 \cdot \varepsilon_\Delta]^{1/2}$. The effective index $n_{eff,(m)} = [\varepsilon_b + (\alpha/k_0)^2 \cdot (p - m)^2]^{1/2} \approx \varepsilon_b^{1/2} \cdot [1 + (\alpha/k_0)^2 \cdot (p - m)^2/(2\varepsilon_b)]$ has level separation varying linearly as $n_{eff,(m)} - n_{eff,(m+1)} = (\alpha/k_0)^2 \cdot (p - m - 1/2)/\varepsilon_b^{1/2}$. The permittivity distribution and the corresponding thickness distribution of the SUSY partner waveguide are then obtained by the application of the intertwining operator $A = (1/k_0) \cdot [\partial_x + p \cdot \alpha \cdot tanh(\alpha x)]$.

Figures 2.19a, b show examples of the multilevel switching for different values of refractive index modulations $\Delta n = 9 \times 10^{-3}$ and $\Delta n = 2.1 \times 10^{-2}$, respectively. The degree of "true" is determined by the transparent eigenmode determined by Δn. Confirming the transparency of each eigenmode *in sequence*, the mode-dependent transmission derives the photonic realization of the membership function in many-valued fuzzy logics (Ross et al. 2009; Zadeh 1997). The SUSY Pöschl-Teller pair then operates as the building block of fuzzy logic systems, with complete isolation at the off state and the use of optical bandwidth.

Ultrafast Building Block for Random Waves The key property of the SUSY-induced collective switching of multimodes is phase independence and complete

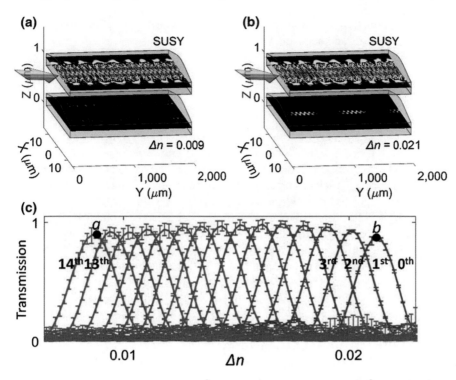

Fig. 2.19 **a** Fourteenth true ($\Delta n = 9 \times 10^{-3}$) and **b** zeroth true ($\Delta n = 2.1 \times 10^{-2}$) transparency for random wave incidences. **c** The transmission of each eigenmode (0th–14th) in the output waveguide at the full coupling position as a function of the modulation Δn. Error bars denote the standard deviation for the ensembles of 400 random samples. Reprinted figure with permission from (Yu et al. 2017). Copyright 2017 by the APS

isolation using the parity-reversal contact designed by the unbroken SUSY transformation. This SUSY-based digital random wave switching boosts the performance of multichannel signal processing in chip-to-chip photonic circuits (Bruck et al. 2015; Dai et al. 2012; Sun et al. 2015) by replacing multi-to-single mode couplers and multiple single mode switches. The necessary level of the refractive index modulation depth $\Delta n/n \sim 0.23\%$ for the designed structure can be achieved with 10–40 Gbps speed, using the carrier density control in silicon, germanium, and their hybrid structures (Reed et al. 2010). We also envisage the use of all-optical modulations using optical nonlinearity (Nozaki et al. 2010; Yu et al. 2012; Albert et al. 2011; Kollmann et al. 2014) by decreasing the level separation in eigenspectra for ultrafast switching applications.

Although we have used the SUSY transformation, in the k-domain in photonics, the same concept can be extended to other wave quantities, such as the ω-domain. For example, the isospectrality and harmonic realizations are directly applicable to the ω-domain to handle temporally random waveforms by using coupled resonators (Hamel et al. 2015) for multiresonant structures. Because of the generality of the

Schrödinger equation, this concept and approach can also be extended to quantum-mechanical, elastic, and acoustic waves.

2.5   Isospectrality by the Householder Transformation

There have been extensive efforts to understand physical (Bianconi and Barabási 2001; Goodrich et al. 2014; Viciani et al. 2015) and biological (Li et al. 2004) complex systems from the perspective of abstract graph networks, whose vertices and edges describe the self- and interacting-dynamics of local elements, respectively. The network analysis of *C. elegans* (Li et al. 2004) with "neuron vertices" and "synapse edges" has established the origin of efficient signal transport despite the low connection density, i.e., "small-world" networks (Watts and Strogatz 1998). In physics, the network-based description of electrical and optical potentials has recently deepened the understanding of "disorder", extending a classic milestone of Anderson localization (Anderson 1958); the extreme form of disordered media has been revealed from the behaviour of interacting particle networks near the jamming transition (Goodrich et al. 2014), and the correlated disorder has paved the way for the observation of non-Anderson-like waves, such as disorder-resistive delocalization (Yu et al. 2016a, b, 2018) or flatband (Bodyfelt et al. 2014).

By examining the relationship between wave physics and its network-based description, we begin with the question, "Can materials in real space cover all the regimes of a general graph network (Fig. 2.20)?" In the graph, you can easily imagine the interaction between far-off vertices just by drawing the edge between the vertices. In real space, however, the strong connection of far-off wave elements is extremely difficult to achieve because local modes that are generated by oscillating fundamental particles have strong spatial dependence that restricts the presence of 'edges' (coupling) between 'vertices' (local modes). For example, electromagnetic point-source resonances have power decay, such as $1/r^2$ (far-field) and $e^{-\alpha r}$ (near-field), that prohibits the design of optical materials with stronger far-off coupling than nearest-neighbour coupling.

As an answer to the above question, in this section, we focus on the real-space photonic reproduction of the level statistics in highly complex "hypothetical" networks. Because these identical level statistics between complex graph networks and

Fig. 2.20   Can we always obtain the photonic structure corresponding to an arbitrary graph network in terms of wave parameters (e.g., spectral information)?

photonic structures directly correspond to the isospectrality, we need to apply the proper isospectral transformation for this problem, here, by exploiting the mathematical similarity between Hamiltonians in different dimensions using the Householder transformation, which allows the photonic realization of a hypothetical network including even significant coupling between far-off vertices. Let us start with the introduction of isospectrality based on the Householder transformation.

Isospectrality between Network and Photonic Structures We start from the interdisciplinary viewpoint of network theory on the design of photonic structures. The light that flows inside structures composed of coupled elements can be interpreted as signal transport over graph networks (Bodyfelt et al. 2014; Lidorikis et al. 1998; Liang and Chong 2013). Reciprocity in electromagnetics constructs the undirected network (Bodyfelt et al. 2014; West and Prentice Hall 2001), whose vertices and edges denote resonances and coupling, respectively. The strength of disorder corresponds to the graph irregularity, and the dimensions of the material structures determine the "degree" of the graphs, which quantifies the number of coupling paths in space. This graph-based perspective reveals that optical structures in 3D real space cover only the restricted parts of general graph theory (West and Prentice Hall 2001), originating from the difficulty in connecting far-off vertices in real space. At most, nearest-neighbour and next-nearest-neighbour coupling (Keil et al. 2015) significantly affect light transport in weakly-coupled systems.

Now, consider the example of a "hypothetical" optical network (Fig. 2.21a) and the corresponding graph representation (Fig. 2.21b). The network is obtained by the Watts-Strogatz rewiring (Watts and Strogatz 1998) of a regular graph; each edge in the graph network that is selected in turn is reconnected with the rewiring probability p to a vertex uniformly chosen at random (number of vertices $N = 21$, graph degree $k = 4$, and rewiring probability $p = 0.5$). Each vertex represents an optical local mode (Lidorikis et al. 1998; Liang and Chong 2013; Garanovich et al. 2012), whereas the edge between vertices denotes a "hypothetical" coupling between elements, including significant long-range coupling. As an illustrative case, we begin with the non-weighted network (West and Prentice Hall 2001), whose edges (or vertices) have equal coupling strength (or resonance). The proposed N-body system is actually unrealistic because the long-range coupling between far-off elements is suppressed for the local modes.

The level statistics of the proposed N-body graph is determined by the eigenvalue equation $H\psi = E\psi$ (Lidorikis et al. 1998; Garanovich et al. 2012) of

$$\frac{d}{dt}\psi_v = i\rho_0\psi_v + \sum_{w \neq v} i\kappa(v, w) \cdot \psi_w, \qquad (2.13)$$

where $v = 1, 2, \ldots, N$ is the vertex number, ψ_v is the electromagnetic field at the vth element, ρ_0 is the self-energy of each element, and $\kappa(v, w)$ is the coupling coefficient between the vth and wth elements for the randomly-rewired graph ($\kappa(v, w) = \kappa_0$ for connected v-w and $\kappa(v, w) = 0$ for disconnected v-w vertices). The

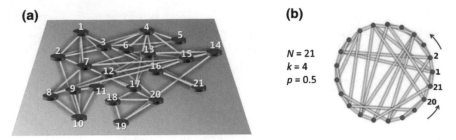

Fig. 2.21 Network-inspired design of 1D structure with a 'similar' Hamiltonian. **a** A schematic of an optical network ($N = 21$, $k = 4$, and $p = 0.5$) composed of hypothetically coupled resonances. **b** The graph representation of the structure. The coupling coefficient is $\kappa_0 = 0.01 \cdot \rho_0$, where the local resonance is normalized as $\rho_0 = 1$ Figure and caption adapted from Yu et al. (2016a, b) under the Gold Open Access Publishing policy of The Optical Society (OSA) (https://www.osapublishing.org/optica/submit/review/open-access-policy-statement.cfm)

system Hamiltonian H is symmetric ($H = H^T$) due to the reciprocity $\kappa(v, w) = \kappa(w, v)$.

In the Watts-Strogatz-rewired (Watts and Strogatz 1998) high-degree graph, the off-diagonal part of its Hamiltonian H is randomly filled with κ_0 or zero, in contrast to the sparse and quasi-diagonal Hamiltonians of real-space structures. To transform this hypothetical H into a realistic form, we focus on the derivation of the effective Hamiltonian, which includes only nearest-neighbour couplings for a 1D structure but possesses identical level statistics. For this purpose, we apply the similarity between symmetric Hamiltonians based on Householder transformation (Kincaid and Cheney 2002). For the orthogonal matrix $M^T M = I$, the eigenvalue equation can be cast in the form of $MH(M^T M)\psi = M(E\psi)$, deriving the isospectral transformation as $H_i(M\psi) = (MHM^T) \cdot (M\psi) = E(M\psi)$, where $H_i = MHM^T$ is the transformed Hamiltonian with an identical eigenspectrum E and transformed eigenmodes $M\psi$. This condition also satisfies the intertwining relation $MH = H_iM$. By applying the tridiagonalization technique (Kincaid 2002) from the Householder transformation, the sequential transformation of $H_{iq} = (M_q M_{q-1} \cdots M_2 M_1) \cdot H \cdot (M_q M_{q-1} \cdots M_2 M_1)^T$ can remove the off-diagonal terms of H in order except the tridiagonal components (Fig. 2.22a) when M_q has the form of

$$M_q = \begin{pmatrix} I_q & O_q \\ O_q & L_{N-q} \end{pmatrix}, \tag{2.14}$$

where I_q and O_q denote the $q \times q$ identity and the null matrix, respectively, L_{N-q} has the Householder form (Kincaid and Cheney 2002) of $L_{N-q} = I_{N-q} - 2uu^T$, and u is the unit normal vector that defines the reflection plane for the Householder transformation. Figure 2.22b shows the serial reduction of the graph edges from the sequential transformation. The transformation of $M = M_{N-2}M_{N-3} \cdots M_2M_1$ ultimately derives the effectively 'weighted' graph with the degree $k = 2$, corresponding to a finite 1D structure in real space with the tridiagonalized Hamiltonian $H_i = MHM^T$ (Fig. 2.22c).

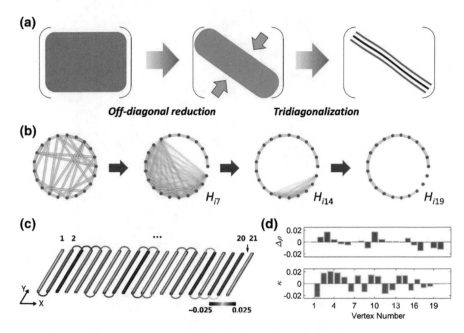

Fig. 2.22 **a** Evolution of the Hamiltonian matrix by applying tridiagonalization, which preserves the original eigenspectrum but sequentially removes off-diagonal terms. **b** The variation of the network based on the sequential Householder transformation. **c** A 1D structure satisfying the Hamiltonian H_{i19} in **b**. The colours of the connections and elements represent the value of κ and the modification of local resonances $\Delta\rho$, respectively. The distributions of $\Delta\rho$ and κ are shown in **d**. $\kappa_0 = 0.01 \cdot \rho_0$, and $\rho_0 = 1$ Figure **b–d** and caption adapted from Yu et al. (2016a, b) under the Gold Open Access Publishing policy of OSA (https://www.osapublishing.org/optica/submit/review/open-access-policy-statement.cfm)

Thus, we show that a graph network, even a graph network that includes significant long-range coupling, has a 1D isospectral partner structure that includes only nearest-neighbour couplings (Fig. 2.22c from the final graph of Fig. 2.22b; see Fig. 2.22d for the corresponding self-energy and coupling strength distribution). However, achieving this isospectral 1D realization of a network does not guarantee that the spectrum of the network can be fully reproduced by a "single" 1D structure, as shown in the disconnection in Fig. 2.22c (black arrow).

From Order to Randomness Because of the difficulty regarding the removal of crosstalk (Mrejen et al. 2015) inside integrated wave structures, ideally, the transformed tridiagonal Hamiltonian H_i needs to be achieved with a single connected structure. The origin of the disconnection in the 1D design can be elucidated by investigating the graph networks of different degrees of disorder. Figure 2.23 shows three rewired graphs with different values of rewiring probability p and their level statics and transformed 1D designs. For the regular graph of $p = 0$, multilevel edges allow non-unique paths towards effective interactions over the whole network that derive energy level degeneracy d (Fig. 2.23b), contrary to the structures only with

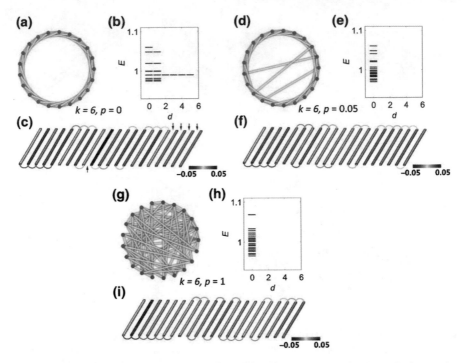

Fig. 2.23 Degeneracy lifted by disorder. **a–c** Regular, **d–f** $p = 0.05$, and **g–i** $p = 1$ rewired networks and the corresponding **a, d, g** graphs, **b, e, h**, level statistics as a function of the degeneracy d, and **c, f, i**, transformed 1D structures Figure and caption adapted from Yu et al. (2016a, b) under the Gold Open Access Publishing policy of OSA (https://www.osapublishing.org/optica/submit/review/open-access-policy-statement.cfm)

nearest-neighbour couplings (Garanovich et al. 2012). This degeneracy originates from the presence of edge-induced symmetry in the graph. Because the finite 1D structure from the tridiagonal H_i lacks enough symmetry for degeneracy of the regular network, a disconnection in the 1D structure is inevitable (black arrows in Fig. 2.23c), which prohibits the existence of a single and isospectral 1D design. We then impose disorder on the regular network (Watts and Strogatz 1998) to break the structural symmetry from the emergence of shortcuts and thus dramatically lift the level degeneracy. Most of the networks in the regime between weak (Fig. 2.23d–f) and completely random (Fig. 2.23g–i) disorder thus possess an isospectral pair in the fully connected 1D structure.

Therefore, we emphasize that the level statistics of hypothetical or unrealistic 'disordered' systems can be fully achieved in a 'single' realistic 1D structure of only nearest-neighbour evanescent couplings through the Householder Hamiltonian transformation $H_i = MHM^T$. Because the regime of strong disorder not only more thoroughly confirms the connected 1D structure but also more clearly shows the broadband nature of disorder (Fig. 2.23h vs. Fig. 2.23e regarding the disorder-induced removal of bandgaps (Yu et al. 2015a, b)), a broadband device based on disordered structures (Garanovich et al. 2012; Moreau et al. 2012; Vynck et al. 2012; Aeschlimann et al. 2015; Hedayati et al. 2011; Redding et al. 2013) is a suitable application of our network-inspired design methodology.

2.6 Application: Interdimensional Wave Transport

How can we use the 1:1 spectral correspondence between network and photonic structures? The key is that the "dimension" of the structure corresponds to the "degree" of the counterpart graph. Therefore, a higher-dimensional (2D, 3D) structure with a disordered coupling network is a representative example of a high-degree random-walk network, which enables spectral correspondence, or isospectrality, between different dimensional photonic structures. **The detailed process is as follows (P-for process):**

P-I. Modelling the 2D/3D photonic structures with the graph network by assigning each optical "element" as a "graph node" and electromagnetic "coupling" between elements as a "graph link".

P-II. Tridiagonalizing the graph network using the Householder transformation, and

P-II. Finding the 1D photonic structure that is the isospectral counterpart of the transformed graph.

As an example, we seek to find the isospectrality between 1D and 3D structures. In detail, we apply our similarity-based Hamiltonian transformation to a real-space design (Fig. 2.24), the transformation of a 3D structure that includes all orders of couplings (with higher-order coupling between far-off elements, (Keil et al. 2015)) into a 1D structure that uses only nearest-neighbour coupling. Figure 2.24a shows the finite-size optical cubic lattice, which possesses periodicity along all the Cartesian axes. The disordered structure (Fig. 2.24b) for the lifted degeneracy is obtained from the 3D random-walk displacement of each atom. Owing to the spatial variation of the coupling κ, the corresponding graph for each structure should have $N(N-1)/2$ "weighted" edges (Fig. 2.24c, d), which are represented by the varying thicknesses of the lines. The periodicity in the x-, y-, and z-axes in the cubic lattice leads to three symmetric edges with identical coupling strengths (yellow lines in Fig. 2.24a), which is the origin of the degenerate states. Thus, the transformed 1D structure of the cubic lattice is divided into three parts ($\kappa = 0$ in Fig. 2.24e), while the random-walk deformed structure with a randomly weighted graph (Fig. 2.24d) derives the fully connected isospectral 1D structure (Fig. 2.24f).

Interdimensional Wave Transport From this "dimensional isospectrality", we demonstrate the lossless energy transfer between different dimensional structures. The isospectrality in multimode structures provides the global phase matching condition, allowing power transfer for all the multimodes. As an application example, we show the multimode coupling between 2D and 1D structures. Figure 2.25a shows

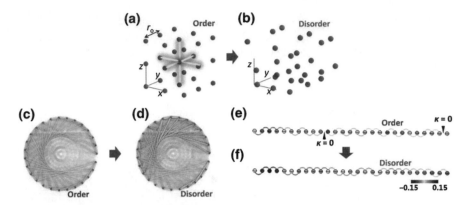

Fig. 2.24 Network-inspired design in real space. **a** A cubic lattice and **b** the corresponding deformed structure. The yellow lines in **a** denote degenerate interactions along the x-, y-, and z-axes, the origin of the degeneracy. For optical 'atoms', following the exponential relation between the coupling κ and the distance between elements r, we use the relation $\kappa = 0.236 \cdot exp(-0.451 \cdot r)$. **c, d** Graph representations and **e, f** transformed 1D structures are shown for the 3D ordered (**c, e**) and disordered (**d, f**) structures in **a, b**. The original periodicity is $r_o = 5$, the local resonance in 3D structures is $\rho_o = 1$, and the location of each element is deformed by $\Delta_{x, y, z} = 2 \cdot unif(-1, 1)$ where $unif(-1, 1)$ is the uniform distribution between -1 and 1 Figure and caption adapted from (Yu et al. 2016a, b) under the Gold Open Access Publishing policy of OSA (https://www.osapublishing.org/optica/submit/review/open-access-policy-statement.cfm)

the proposed structure. From the original 2D structure composed of identical optical atoms (blue dashed circle), we achieve the 1D isospectral partner structure (red dashed ellipse) by following the transformation $H_i = MHM^T$ (Fig. 2.25b for the self-energy and coupling strength distribution of the 1D structure and Fig. 2.25c for the identical level statistics of both structures). The 2D and 1D structures are coupled through a single atom of each structure with the coupling coefficient κ_d. To preserve the global phase matching condition, the strength of κ_d is lower than that of the internal couplings in the 2D and 1D structures as $\kappa_d < \kappa_{ij}$. Figure 2.26 shows the coupling between the 2D and 1D structures. As an example, the 3rd and 9th eigenmodes are excited in the 2D structure. Due to the matching of each eigenvalue, full couplings between different dimensions are successfully achieved for both eigenmodes. Figure 2.26c–h presents the temporal evolutions of the field distribution for the excitation of the 3rd and 9th eigenmodes in the 2D structure.

Applications We showed that the similarity between Hamiltonians enables the reduction of network degrees with preserved level statics, thereby realizing interdimensional isospectrality for real-space materials. A 2D/3D 'disordered' structure, even a structure that includes all higher-order couplings, can be transformed into a single 1D structure with nearest-neighbour coupling. From this result, we envisage the ultimate simplification in the design strategy in photonics: experimentally convenient 1D broadband absorption replicating 2D/3D disordered structures in the spectral domain. Furthermore, from the interdimensional global phase matching, efficient

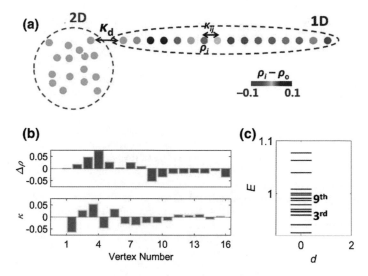

Fig. 2.25 **a** Schematic for 2D and 1D structures. $\kappa = 0.236 \cdot exp(-0.451 \cdot r)$, $r_o = 5$, $\rho_o = 1$, and $\Delta_{x,y,z} = 2 \cdot unif(-1, 1)$. The interdimensional coupling $\kappa_d = 0.01$. The colours of the elements represent the value of the modification of local resonances $\Delta\rho = \rho - \rho_o$. **b** The distributions of $\Delta\rho$ and κ in the 1D structure. **c** The level statistics of both structures Figure and caption adapted from Yu et al. (2016a, b) under the Gold Open Access Publishing policy of OSA (https://www. osapublishing.org/optica/submit/review/open-access-policy-statement.cfm)

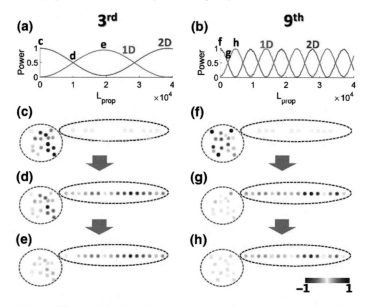

Fig. 2.26 Interdimensional transport between 2D and 1D structures. **a, b** Power flow in each structure for the propagation length L_{prop}. **c–h** Field distributions in both structures at the propagation lengths depicted in (**a, b**). **a, c–e**: 3rd state, **b, f–h**: 9th state Figure and caption adapted from Yu et al. (2016a, b) under the Gold Open Access Publishing policy of OSA (https://www.osapublishing.org/optica/submit/review/open-access-policy-statement.cfm)

wave transfers between different dimensions can be achieved: the perfect energy delivery between 2D optical receptors and 1D guiding systems.

References

Aeschlimann, M., Brixner, T., Differt, D., Heinzmann, U., Hensen, M., Kramer, C., Lükermann, F., Melchior, P., Pfeiffer, W., Piecuch, M.: Perfect absorption in nanotextured thin films via Anderson-localized photon modes. Nat. Photon **9**, 663 (2015)

Albert, M., Dantan, A., Drewsen, M.: Cavity electromagnetically induced transparency and all-optical switching using ion coulomb crystals. Nat. Photon **5**, 633–636 (2011)

Anderson, P.W.: Absence of diffusion in certain random lattices. Phys. Rev **109**, 1492 (1958)

Andrianov, A.A., Borisov, N., Ioffe, M.: The factorization method and quantum systems with equivalent energy spectra. Phys. Lett. A **105**, 19–22 (1984a)

Andrianov, A.A., Borisov, N.V., Ioffe, M.V.: Factorization method and Darboux transformation for multidimensional Hamiltonians. Theoret. Math. Phys **61**, 1078–1088 (1984b)

Bagrov, V.G., Samsonov, B.F.: Darboux transformation, factorization, and supersymmetry in one-dimensional quantum mechanics. Theoret. Math. Phys **104**, 1051–1060 (1995)

Barkhofen, S., Lorz, L., Nitsche, T., Silberhorn, C., Schomerus, H.: Supersymmetric polarization anomaly in photonic discrete-time quantum walks. Phys. Rev. Lett **121**, 260501 (2018)

Bianconi, G., Barabási, A.-L.: Bose-Einstein condensation in complex networks. Phys. Rev. Lett **86**, 5632 (2001)

Bodyfelt, J.D., Leykam, D., Danieli, C., Yu, X., Flach, S.: Flatbands under correlated perturbations. Phys. Rev. Lett **113**, 236403 (2014)

Brossard, J., Carmona, R.: Can one hear the dimension of a fractal? Commun. Math. Phys **104**, 103–122 (1986)

Bruck, R., Mills, B., Troia, B., Thomson, D.J., Gardes, F.Y., Hu, Y., Mashanovich, G.Z., Passaro, V.M., Reed, G.T., Muskens, O.L.: Device-level characterization of the flow of light in integrated photonic circuits using ultrafast photomodulation spectroscopy. Nat. Photon **9**, 54–60 (2015)

Cannata, F., Ioffe, M.V., Nishnianidze, D.: Two-dimensional SUSY-pseudo-hermiticity without separation of variables. Phys. Lett. A **310**, 344–352 (2003)

Chaunsali, R., Xu, H., Yang, J., Kevrekidis, P.: Linear and nonlinear dynamics of isospectral granular chains. Jour. Phys. A **50**, 175201 (2017)

Chumakov, S.M., Wolf, K.B.: Supersymmetry in Helmholtz optics. Phys. Lett. A **193**, 51–53 (1994)

Clark, A.H.: Electrical and optical properties of amorphous germanium. Phys. Rev **154**, 750–757 (1967). https://doi.org/10.1103/PhysRev.154.750

Cooper, F., Khare, A., Sukhatme, U.: Supersymmetry in quantum mechanics. World Scientific (2001)

Dai, D., Bauters, J., Bowers, J.E.: Passive technologies for future large-scale photonic integrated circuits on silicon: polarization handling, light non-reciprocity and loss reduction. Light Sci. Appl **1**, e1 (2012). https://doi.org/10.1038/lsa.2012.1

DiStasio Jr., R.A., Zhang, G., Stillinger, F.H., Torquato, S.: Rational design of stealthy hyperuniform two-phase media with tunable order. Phys. Rev. E **97**, 023311 (2018)

Demircioğlu, B., Kuru, Ş., Önder, M., Vercin, A.: Two families of superintegrable and isospectral potentials in two dimensions. J. Math. Phys **43**, 2133–2150 (2002)

Dreisow, F., Szameit, A., Heinrich, M., Pertsch, T., Nolte, S., Tünnermann, A., Longhi, S.: Bloch-Zener oscillations in binary superlattices. Phys. Rev. Lett **102**, 076802 (2009)

Duplantier, B.: Can one "hear" the thermodynamics of a (rough) colloid? Phys. Rev. Lett **66**, 1555 (1991)

Edagawa, K., Kanoko, S., Notomi, M.: Photonic amorphous diamond structure with a 3D photonic band gap. Phys. Rev. Lett **100**, 013901 (2008)

El-Ganainy, R., Ge, L., Khajavikhan, M., Christodoulides, D.N.: Supersymmetric laser arrays. Phys. Rev. A **92**, 033818 (2015)

El-Ganainy, R., Makris, K.G., Khajavikhan, M., Musslimani, Z.H., Rotter, S., Christodoulides, D.N.: Non-Hermitian physics and PT symmetry. Nat. Phys **14**, 11 (2018)

Ellis, J.: Prospects for Supersymmetry at the LHC & Beyond. arXiv preprint arXiv:1510.06204 (2015)

Feng, L., El-Ganainy, R., Ge, L.: Non-Hermitian photonics based on parity–time symmetry. Nat. Photon **11**, 752–762 (2017)

Florescu, M., Torquato, S., Steinhardt, P.J.: Designer disordered materials with large, complete photonic band gaps. Proc. Natl. Acad. Sci. USA **106**, 20658–20663 (2009)

Florescu, M., Steinhardt, P.J., Torquato, S.: Optical cavities and waveguides in hyperuniform disordered photonic solids. Phys. Rev. B **87**, 165116 (2013)

Ganesh, R., Gonella, S.: From modal mixing to tunable functional switches in nonlinear phononic crystals. Phys. Rev. Lett. **114**, 054302 (2015)

Garanovich, I.L., Longhi, S., Sukhorukov, A.A., Kivshar, Y.S.: Light propagation and localization in modulated photonic lattices and waveguides. Phys. Rep **518**, 1–79 (2012)

Goodrich, C.P., Liu, A.J., Nagel, S.R.: Solids between the mechanical extremes of order and disorder. Nat. Phys **10**, 578 (2014)

Gordon, R.: Harmonic oscillation in a spatially finite array waveguide. Opt. Lett **29**, 2752–2754 (2004)

Gordon, C., Webb, D.L., Wolpert, S.: One cannot hear the shape of a drum. Bull. Am. Math. Soc **27**, 134–138 (1992)

Grover, T., Sheng, D., Vishwanath, A.: Emergent space-time supersymmetry at the boundary of a topological phase. Science **344**, 280–283 (2014)

Gutkin, B., Smilansky, U.: Can one hear the shape of a graph? arXiv preprint nlin/0105020 (2001)

Gutshabash, E. S.: Moutard transformation and its application to some physical problems. I. The case of two independent variables. Journal of Mathematical Sciences 192, 57–69 (2013)

Hamel, P., Haddadi, S., Raineri, F., Monnier, P., Beaudoin, G., Sagnes, I., Levenson, A., Yacomotti, A.M.: Spontaneous mirror-symmetry breaking in coupled photonic-crystal nanolasers. Nat. Photon **9**, 311–315 (2015)

Hedayati, M.K., Javaherirahim, M., Mozooni, B., Abdelaziz, R., Tavassolizadeh, A., Chakravadhanula, V.S.K., Zaporojtchenko, V., Strunkus, T., Faupel, F., Elbahri, M.: Design of a perfect black absorber at visible frequencies using plasmonic metamaterials. Adv. Mater **23**, 5410 (2011)

Heinrich, M., Miri, M.A., Stutzer, S., El-Ganainy, R., Nolte, S., Szameit, A., Christodoulides, D.N.: Supersymmetric mode converters. Nat. Commun **5**, 3698 (2014a). https://doi.org/10.1038/ncomms4698

Heinrich, M., Miri, M.A., Stutzer, S., Nolte, S., Christodoulides, D.N., Szameit, A.: Observation of supersymmetric scattering in photonic lattices. Opt. Lett **39**, 6130–6133 (2014b). https://doi.org/10.1364/OL.39.006130

Hejna, M., Steinhardt, P.J., Torquato, S.: Nearly hyperuniform network models of amorphous silicon. Phys. Rev. B 87 (2013). https://doi.org/10.1103/physrevb.87.245204

Hokmabadi, M.P., Nye, N.S., El-Ganainy, R., Christodoulides, D.N., Khajavikhan, M.: Supersymmetric laser arrays. Science **363**, 623–626 (2019)

Hurst, H.E.: Long-term storage capacity of reservoirs. Trans. Amer. Soc. Civil Eng **116**, 770–808 (1951)

Joannopoulos, J.D., Johnson, S.G., Winn, J.N., Meade, R.D.: Photonic crystals: molding the flow of light. Princeton University Press (2011)

Kac, M.: Can one hear the shape of a drum? Am. Math. Monthly **73**, 1–23 (1966)

Keil, R., Pressl, B., Heilmann, R., Gräfe, M., Weihs, G., Szameit, A.: Direct measurement of second-order coupling in a waveguide lattice. Appl. Phys. Lett **107**, 241104 (2015)

Keilmann, T., Lanzmich, S., McCulloch, I., Roncaglia, M.: Statistically induced phase transitions and anyons in 1D optical lattices. Nat. Commun **2**, 361 (2011)

Kincaid, D.R., Cheney, E.W.: Numerical analysis: mathematics of scientific computing. American Mathematical Society (2002)

King, C., Horsley, S., Philbin, T.: Perfect Transmission through Disordered Media. Phys. Rev. Lett. **118**, 163201 (2017)

Kittel, C., McEuen, P., McEuen, P.: Introduction to solid state physics. Wiley, New York (1976)

Kuru, Ş., Teğmen, A., Vercin, A.: Intertwined isospectral potentials in an arbitrary dimension. J. Math. Phys **42**, 3344–3360 (2001)

Laba, H.P., Tkachuk, V.M.: Quantum-mechanical analogy and supersymmetry of electromagnetic wave modes in planar waveguides. Phys. Rev. A 89 (2014). https://doi.org/10.1103/physreva.89.033826

Levine, D., Steinhardt, P.J.: Quasicrystals: a new class of ordered structures. Phys. Rev. Lett **53**, 2477 (1984)

Li, S., Armstrong, C.M., Bertin, N., Ge, H., Milstein, S., Boxem, M., Vidalain, P.-O., Han, J.-D.J., Chesneau, A., Hao, T.: A map of the interactome network of the metazoan C. elegans. Science 303, 540–543 (2004)

Li, P., Yang, X., Maß, T.W., Hanss, J., Lewin, M., Michel, A.-K.U., Wuttig, M., Taubner, T.: Reversible optical switching of highly confined phonon-polaritons with an ultrathin phase-change material. Nat. Mater **15**, 870–875 (2016)

Liang, G., Chong, Y.: Optical resonator analog of a two-dimensional topological insulator. Phys. Rev. Lett **110**, 203904 (2013)

Lidorikis, E., Sigalas, M., Economou, E.N., Soukoulis, C.: Tight-binding parametrization for photonic band gap materials. Phys. Rev. Lett **81**, 1405 (1998)

Longhi, S.: Quantum-optical analogies using photonic structures. Laser Photon. Rev **3**, 243–261 (2009)

Longhi, S.: Talbot self-imaging in PT-symmetric complex crystals. Phys. Rev. A **90**, 043827 (2014)

Longhi, S.: Supersymmetric transparent optical intersections. Opt. Lett **40**, 463 (2015a)

Longhi, S.: Supersymmetric bragg gratings. J. Opt **17**, 045803 (2015b)

Lykken, J., Spiropulu, M.: Supersymmetry and the crisis in physics. Sci. Am **310**, 34–39 (2014)

Macho, A., Llorente, R., García-Meca, C.: Supersymmetric transformations in optical fibers. Phys. Rev. Appl **9**, 014024 (2018)

Man, W., Florescu, M., Williamson, E.P., He, Y., Hashemizad, S.R., Leung, B.Y., Liner, D.R., Torquato, S., Chaikin, P.M., Steinhardt, P.J.: Isotropic band gaps and freeform waveguides observed in hyperuniform disordered photonic solids. Proc. Natl. Acad. Sci **110**, 15886–15891 (2013)

Mandelbrot, B.B., Wallis, J.R.: Robustness of the rescaled range R/S in the measurement of non-cyclic long run statistical dependence. Water Resour. Res **5**, 967–988 (1969)

Marder, M.P.: Condensed matter physics. John Wiley & Sons (2010)

Midya, B., Walasik, W., Litchinitser, N.M., Feng, L.: Supercharge optical arrays. Opt. Lett **43**, 4927–4930 (2018)

Miri, M.-A., Heinrich, M., El-Ganainy, R., Christodoulides, D.N.: Supersymmetric optical structures. Phys. Rev. Lett **110**, 233902 (2013a)

Miri, M.-A., Heinrich, M., Christodoulides, D.N.: Supersymmetry-generated complex optical potentials with real spectra. Phys. Rev. A **87**, 043819 (2013b)

Miri, M.-A., Heinrich, M., Christodoulides, D.N.: SUSY-inspired one-dimensional transformation optics. Optica **1**, 89 (2014). https://doi.org/10.1364/optica.1.000089

Moon, C.R., Mattos, L.S., Foster, B.K., Zeltzer, G., Ko, W., Manoharan, H.C.: Quantum phase extraction in isospectral electronic nanostructures. Science **319**, 782–787 (2008)

Moreau, A., Ciracì, C., Mock, J.J., Hill, R.T., Wang, Q., Wiley, B.J., Chilkoti, A., Smith, D.R.: Controlled-reflectance surfaces with film-coupled colloidal nanoantennas. Nature **492**, 86 (2012)

Mrejen, M., Suchowski, H., Hatakeyama, T., Wu, C., Feng, L., O'Brien, K., Wang, Y., Zhang, X.: Adiabatic elimination-based coupling control in densely packed subwavelength waveguides. Nat. Commun **6**, 7565 (2015)

Nakata, Y., Urade, Y., Nakanishi, T., Miyamaru, F., Takeda, M.W., Kitano, M.: Supersymmetric correspondence in spectra on a graph and its line graph: From circuit theory to spoof plasmons on metallic lattices. Phys. Rev. A **93**, 043853 (2016)

Nozaki, K., Tanabe, T., Shinya, A., Matsuo, S., Sato, T., Taniyama, H., Notomi, M.: Sub-femtojoule all-optical switching using a photonic-crystal nanocavity. Nat. Photon **4**, 477–483 (2010)

Piao, X., Yu, S., Koo, S., Lee, K., Park, N.: Fano-type spectral asymmetry and its control for plasmonic metal-insulator-metal stub structures. Opt. Express **19**, 10907–10912 (2011)

Piao, X., Yu, S., Park, N.: Control of Fano asymmetry in plasmon induced transparency and its application to plasmonic waveguide modulator. Opt. Express **20**, 18994–18999 (2012)

Piao, X., Yu, S., Hong, J., Park, N.: Spectral separation of optical spin based on antisymmetric Fano resonances. Sci. Rep **5**, 16585 (2015). https://doi.org/10.1038/srep16585

Principe, M., Castaldi, G., Consales, M., Cusano, A., Galdi, V.: Supersymmetry-inspired non-Hermitian optical couplers. Sci. Rep **5**, 8568 (2015). https://doi.org/10.1038/srep08568

Queraltó, G., Ahufinger, V., Mompart, J.: Integrated photonic devices based on adiabatic transitions between supersymmetric structures. Opt. Express **26**, 33797–33806 (2018)

Ramond, P.: Dual theory for free fermions. Phys. Rev. D **3**, 2415 (1971)

Rechtsman, M., Szameit, A., Dreisow, F., Heinrich, M., Keil, R., Nolte, S., Segev, M.: Amorphous photonic lattices: band gaps, effective mass, and suppressed transport. Phys. Rev. Lett **106**, 193904 (2011)

Redding, B., Liew, S.F., Sarma, R., Cao, H.: Compact spectrometer based on a disordered photonic chip. Nat. Photon **7**, 746 (2013)

Reed, G.T., Mashanovich, G., Gardes, F., Thomson, D.: Silicon optical modulators. Nat. Photon **4**, 518–526 (2010)

Roche, S., Bicout, D., Maciá, E., Kats, E.: Long range correlations in DNA: scaling properties and charge transfer efficiency. Phys. Rev. Lett **91**, 228101 (2003)

Ross, T.J.: Fuzzy logic with engineering applications. John Wiley & Sons (2009)

Sakurai, J.J., Tuan, S.-F., Commins, E.D.: Modern quantum mechanics. AAPT (1995)

Sapienza, R., Costantino, P., Wiersma, D., Ghulinyan, M., Oton, C.J., Pavesi, L.: Optical analogue of electronic bloch oscillations. Phys. Rev. Lett **91**, 263902 (2003)

Schindler, J., Li, A., Zheng, M.C., Ellis, F.M., Kottos, T.: Experimental study of active LRC circuits with PT symmetries. Phys. Rev. A **84**, 040101 (2011)

Schmidt, B., Friedrich, B.: Supersymmetry and eigensurface topology of the planar quantum pendulum. Front. Phys. **2** (2014). https://doi.org/10.3389/fphy.2014.00037

Schwartz, T., Bartal, G., Fishman, S., Segev, M.: Transport and Anderson localization in disordered two-dimensional photonic lattices. Nature **446**, 52–55 (2007)

Smirnova, D.A., Padmanabhan, P., Leykam, D.: Parity anomaly laser. Opt. Lett **44**, 1120–1123 (2019)

Streetman, B.G., Banerjee, S.: Solid state electronic devices. 5th edn, Prentice Hall New Jersey (2000)

Sun, C., Wade, M.T., Lee, Y., Orcutt, J.S., Alloatti, L., Georgas, M.S., Waterman, A.S., Shainline, J.M., Avizienis, R.R., Lin, S.: Single-chip microprocessor that communicates directly using light. Nature **528**, 534–538 (2015)

Teimourpour, M., Christodoulides, D.N., El-Ganainy, R.: Optical revivals in nonuniform supersymmetric photonic arrays. Opt. Lett **41**, 372–375 (2016)

Tissier, M., Tarjus, G.: Supersymmetry and its spontaneous breaking in the random field ising model. Phys. Rev. Lett. **107** (2011). https://doi.org/10.1103/physrevlett.107.041601

Torquato, S.: Hyperuniformity and its generalizations. Phys. Rev. E **94**, 022122 (2016)

Torquato, S., Stillinger, F.H.: Local density fluctuations, hyperuniformity, and order metrics. Phys. Rev. E **68**, 041113 (2003)

Vardeny, Z.V., Nahata, A., Agrawal, A.: Optics of photonic quasicrystals. Nat. Photon **7**, 177–187 (2013)

Viciani, S., Lima, M., Bellini, M., Caruso, F.: Observation of noise-assisted transport in an all-optical cavity-based network. Phys. Rev. Lett **115**, 083601 (2015)

Vynck, K., Burresi, M., Riboli, F., Wiersma, D.S.: Photon management in two-dimensional disordered media. Nat. Mater **11**, 1017 (2012)

Walasik, W., Midya, B., Feng, L., Litchinitser, N.M.: Supersymmetry-guided method for mode selection and optimization in coupled systems. Opt. Lett **43**, 3758–3761 (2018)

Watts, D.J., Strogatz, S.H.: Collective dynamics of 'small-world' networks. Nature **393**, 440–442 (1998)

Weaire, D., Thorpe, M.: Electronic properties of an amorphous solid. I. A simple tight-binding theory. Phys. Rev. B 4, 2508 (1971)

West, D.B.: Introduction to graph theory. Prentice Hall (2001)

Wiersma, D.S.: Disordered photonics. Nat. Photon **7**, 188–196 (2013)

Witten, E.: Dynamical breaking of supersymmetry. Nucl. Phys. B **188**, 513–554 (1981)

Woit, P.: Not even wrong: The failure of string theory and the continuing challenge to unify the laws of physics. Random House (2011)

Wright, L.G., Wabnitz, S., Christodoulides, D.N., Wise, F.W.: Ultrabroadband dispersive radiation by spatiotemporal oscillation of multimode waves. Phys. Rev. Lett **115**, 223902 (2015)

Yu, S., Koo, S., Park, N.: Coded output photonic A/D converter based on photonic crystal slow-light structures. Opt. Express **16**, 13752–13757 (2008)

Yu, S., Piao, X., Koo, S., Shin, J.H., Lee, S.H., Min, B., Park, N.: Mode junction photonics with a symmetry-breaking arrangement of mode-orthogonal heterostructures. Opt. Express **19**, 25500–25511 (2011). https://doi.org/10.1364/OE.19.025500

Yu, S., Piao, X., Park, N.: Slow-light dispersion properties of multiatomic multiband coupled-resonator optical waveguides. Phys. Rev. A **85**, 023823 (2012)

Yu, S., Mason, D.R., Piao, X., Park, N.: Phase-dependent reversible nonreciprocity in complex metamolecules. Phys. Rev. B **87**, 125143 (2013)

Yu, S., Piao, X., Hong, J., Park, N.: Bloch-like waves in random-walk potentials based on supersymmetry. Nat. Commun **6**, 8269 (2015a)

Yu, S., Piao, X., Hong, J., Park, N.: Progress toward high-Q perfect absorption: A fano antilaser. Phys. Rev. A **92**, 011802 (2015b)

Yu, S., Piao, X., Hong, J., Park, N.: Metadisorder for designer light in random systems. Sci. Adv **2**, e1501851 (2016a)

Yu, S., Piao, X., Hong, J., Park, N.: Interdimensional optical isospectrality inspired by graph networks. Optica **3**, 836–839 (2016b)

Yu, S., Piao, X., Park, N.: Controlling random waves with digital building blocks based on supersymmetry. Phys. Rev. Appl. **8**, 054010 (2017)

Yu, S., Piao, X., Park, N.: Disordered Potential landscapes for anomalous delocalization and superdiffusion of light. ACS Photon **5**, 1499 (2018)

Zadeh, L.A.: Toward a theory of fuzzy information granulation and its centrality in human reasoning and fuzzy logic. Fuzzy Set. Syst **90**, 111–127 (1997)

Zaitsev, V.F., Polyanin, A.D.: Handbook of exact solutions for ordinary differential equations. CRC Press (2002)

Zhu, X., Ramezani, H., Shi, C., Zhu, J., Zhang, X.: PT-symmetric acoustics. Phys. Rev. X **4**, 031042 (2014)

Zuniga-Segundo, A., Rodriguez-Lara, B.M., Fernandez, C.D., Moya-Cessa, H.M.: Jacobi photonic lattices and their SUSY partners. Opt. Express **22**, 987–994 (2014). https://doi.org/10.1364/OE. 22.000987

Chapter 3
Designing Modes in Disordered Photonic Structures

Abstract In this chapter, we introduce the inverse design of disordered photonic structures with the target modal response. Starting from the concept of metadisorder, which is defined by order-random interactions in the multivariable parameter space, we discuss the realization of anomalous wave transport and localization in photonic structures as an example of the independent control of wave quantities. As a generalization of this approach, we introduce the concept of Bohmian photonics inspired by the alternative interpretation of quantum mechanics, which enables independent control of the amplitude and phase information of light. Various practical applications of Bohmian photonics, including optical phase trapping, phase-conserved energy confinement and annihilation, and constant-intensity wave transport, are discussed.

Keywords Metadisorder · Anderson localization · Wave transport · Coupled resonators · Chirality · Cloaking · Non-Hermitian photonics · Parity-time symmetry · Bohmian photonics

3.1 Concept of Metadisorder

In Chap. 2, we explored the design of "eigenspectra" in disordered structures, successfully achieving the perfect bandgap with modal localization, random wave switching for binary and fuzzy logics, and lossless signal transport between different dimensional structures. When we consider the eigensystem of light waves, the next natural step is the design of "eigenmodes" in disordered photonic structures. **We raise certain questions (Q-) for this purpose:**

Q-I. Can we achieve well-organized wave transport inside highly disordered structures (Fig. 3.1)?

Q-II. Can we design a seemingly contradictory photonic structure that allows crystal-like modal properties despite random-like spectral information?

© The Author(s), under exclusive license to Springer Nature Singapore Pte. Ltd. 2019   47
S. Yu et al., *Top-Down Design of Disordered Photonic Structures*,
SpringerBriefs in Physics, https://doi.org/10.1007/978-981-13-7527-9_3

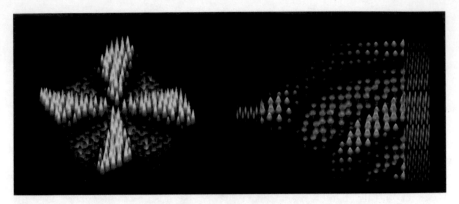

Fig. 3.1 Is it possible to achieve such well-organized (e.g., coherent, directive, and functional) wave phenomena in highly disordered photonic structures?

To answer these questions, first, we examine the properties of wave transport inside random potentials. In his ground-breaking paper (Anderson 1958), P. W. Anderson found that disordered potentials possess "localized" modes, known as Anderson localization, that significantly suppress wave transport. After this finding, it has been strongly believed that the degree of disorder ultimately determines the degree of localization (Schwartz et al. 2007). Thus, Anderson localization has been considered to be a critical indicator that defines the "order" and "disorder" in wave phenomena, resulting in a clear distinction between (i) transporting devices using ordered potentials and (ii) focusing devices using disordered potentials. Previous approaches have mainly focused on filling the gap between order and disorder, wherein the increase of disorder has led to the monotonous change from extended to localized states (Schwartz et al. 2007). However, this classic viewpoint, based on a *single* indicator of order and randomness, may conceal the presence of anomalous transport and localization properties in highly disordered potentials.

To gain insight, let us revisit disorder in another field, network theory, which was discussed in Sect. 1.1. In the pioneering work examining randomly rewired connections in graphs (Watts and Strogatz 1998a), D. J. Watts first discovered the existence of "small-world" disordered graphs between regular and random graphs. This intermediate regime simultaneously allows efficient and robust signal transport and can model various elemental systems in physics, biology, and sociology, such as seismic networks, *C. elegans* neurons (Amaral et al. 2000), brain connectomes (Bullmore and Bassett 2011; Hilgetag and Goulas 2016), and affinity groups in social networks (Watts and Strogatz 1998a). As noted in Sect. 1.1, the small-world regime is defined by the discrepancy between two statistical quantities, the path length and clustering coefficient, in contrast to the corresponding synchronized evolutions in regular and random graphs. Therefore, the intermediate regime between order and randomness can be defined in different ways depending on the choice of the state indicator (path length or clustering coefficient) of the graph structure. Thus, the discovery of the small-world regime demonstrates that the use of proper "multiple" indicators is essential to classify the intermediate regime between order and randomness, which can resolve the inherent complexity and ambiguity of disorder.

In similar context, in photonics, we can envisage the existence of anomalous intermediate regimes between order and randomness, which are defined with multiple wave quantities, e.g., random-like modal "localization" with ordered, well-organized wave "transport" (for the question Q-I). Then, to handle these multiple quantities, we need to quantify photonic structures with multivariable parameters (for the question Q-II) such as on-site energy and hopping energy for weakly coupled systems, in-plane structural parameters and vertical parameter for 3D platforms, and spatial density and correlation in randomly distributed photonic atoms.

Definition of Metadisorder From this approach, we develop the concept of "Metadisorder", as an analogy for the term "metamaterial", which refers to an artificially designed material (**D- for definition**):

D-I. Multiple structural parameters are introduced to quantify an optical potential.
D-II. We selectively impose order and randomness on these parameters.
D-III. The interactions between the structural order and randomness are achieved through *"waves"* from light-matter interactions, allowing anomalous wave phenomena in disordered photonics, e.g., crystal-like modal properties in highly disordered potentials.

As an example of this concept of metadisorder, we examine randomly coupled optical systems that enable globally collective and delocalized waves in highly disordered potentials.

3.2 Metadisorder Configuration in Coupled Systems

As an example of metadisorder platforms, we consider random systems composed of coupled elements (Fig. 3.2), such as waveguides (Garanovich et al. 2012; Christodoulides et al. 2003) (Fig. 3.2a, c) and resonators (Liang and Chong 2013; Lidorikis et al. 1998) (Fig. 3.2b, d). Such coupled systems include two structural degrees of freedom (DOFs), on-site energy and hopping energy. The on-site energy is determined by the wavevector or resonant frequency of each waveguide or resonator, respectively, and can be controlled by the size, shape, and material parameters of each element. The hopping energy is determined by the coupling coefficient between elements. Because we assume randomly coupled systems, the spatial distribution of the hopping energy is defined by the initial random distribution of optical elements.

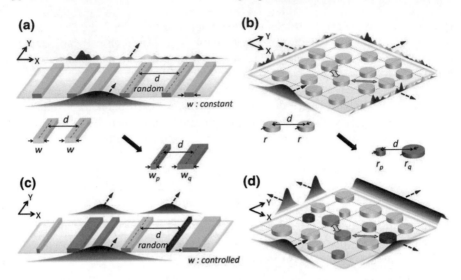

Fig. 3.2 Metadisorder coupled systems. **a**, **b** Illustrate random scattering and Anderson localization in randomly coupled systems composed of identical **a** waveguides or **b** resonators. **c**, **d** Show delocalized propagations of designer modal profiles in metadisorder coupled systems. The control of self-energy distributions can be obtained by manipulating **c** waveguide widths $w_{p,q}$ or **d** resonator radii $r_{p,q}$. The positions of optical elements in coupled systems are assumed to be completely random. Blue dotted arrows denote propagation directions. Figure and caption adapted from Yu et al. (2016a) under American Association for the Advancement of Science (AAAS)'s License to Publish (http://www.sciencemag.org/help/reprints-and-permissions)

The expected wave phenomena inside metadisorder coupled systems are as follows; in contrast to random scattering, incoherence, and Anderson localization in classic random systems composed of identical optical elements (Fig. 3.2a, b), well-organized, coherent, and functional wave phenomena can be obtained in metadisorder coupled systems by the designed distribution of on-site energy (Fig. 3.2c, d). The metadisorder structures also allow ordered waves with disorder-like spectral responses. Such anomalous responses can be understood as the results of the interactions between structural order (on-site energy) and randomness (hopping energy) through "waves".

Analysis of Coupled Systems We consider the analysis of an N-body system composed of weakly coupled optical elements, which can be modelled by the eigenvalue equation $H\psi = \gamma\psi$ using discrete models of coupled-mode theory (CMT) (Garanovich et al. 2012; Haus 1984) or tight-binding (TB) analysis (Lidorikis et al. 1998). For example, the governing equation, including on-site- and hopping-energy, becomes

$$\frac{d}{d\xi}\psi_p = i\gamma_{op}\psi_p + \sum_{q\neq p} i\kappa_{pq}\psi_q, \tag{3.1}$$

where $p = 1, 2, \ldots, N$ is the element number and ψ_p is the field at the pth element; ξ is the evolution axis, which becomes time t for resonators (Joannopoulos et al. 2011) and space x, y, or z for waveguides (Garanovich et al. 2012); γ_{op} is the on-site energy of the pth element, and κ_{pq} is the coupling coefficient for the hopping energy between the pth and qth elements. For the steady state ($\partial_\xi \rightarrow i\gamma$), Eq. (3.1) becomes the matrix eigenvalue equation $H\psi = \gamma\psi$, where

$$
H = \begin{bmatrix} \gamma_{o1} & \kappa_{12} & \cdots & \kappa_{1N} \\ \kappa_{21} & \gamma_{o2} & \cdots & \kappa_{2N} \\ \vdots & \vdots & \ddots & \vdots \\ \kappa_{N1} & \kappa_{N2} & \cdots & \gamma_{oN} \end{bmatrix}
$$

$$
= D + K = \begin{bmatrix} \gamma_{o1} & 0 & \cdots & 0 \\ 0 & \gamma_{o2} & \cdots & 0 \\ \vdots & \vdots & \ddots & \vdots \\ 0 & 0 & \cdots & \gamma_{oN} \end{bmatrix} + \begin{bmatrix} 0 & \kappa_{12} & \cdots & \kappa_{1N} \\ \kappa_{21} & 0 & \cdots & \kappa_{2N} \\ \vdots & \vdots & \ddots & \vdots \\ \kappa_{N1} & \kappa_{N2} & \cdots & 0 \end{bmatrix}
\tag{3.2}
$$

and $\psi = [\psi_1, \psi_2, \ldots, \psi_N]^T$. Because each element number p corresponds to the spatial position of the pth element $\mathbf{X}_p$ (e.g., $p \rightarrow x_p$ for 1D and $p \rightarrow (x_p, y_p)$ for 2D problems), ψ can be re-expressed in the spatial function $\psi = \psi(\mathbf{X})$. Additionally, as shown in Eq. (3.2), the Hamiltonian H can be decomposed as $H = D + K$, where D is the diagonal on-site energy matrix (e.g., wavevector β of each waveguide or resonant frequency f of each resonator, both of which define the phase evolution of the field in uncoupled optical elements (Haus 1984)) and K is the off-diagonal matrix for the hopping energy between elements (e.g., coupling coefficient κ between coupled waveguides or between coupled resonators (Haus 1984)), which represents the network of the system. The randomly coupled system is described by the random matrix K.

Conventional Disorder Figure 3.3a shows an example of conventional photonic disorder, a 1-dimensional (1D) coupled system with off-diagonal disorder (Martin et al. 2011) with identical on-site energy β and random interactions ($[D]_p = \gamma_{o0}$ and $[K]_{pq} = \kappa_{pq} = \kappa_0 + \Delta\kappa \cdot u(-1, 1)$, where κ_{pq} is the coupling between the pth and qth elements for $1 \leq p, q \leq N$, κ_0 and $\Delta\kappa$ represent the averaged and disordered couplings, respectively, and u is the uniform probability density function).

According to the discussion in Sect. 3.1, we focus on two characteristics that describe wave behaviours inside disordered systems, localization and transport, similar to the clustering and path length that are applied to define small-world graphs. Figure 3.3b–d presents a few eigenmodes of each system at different degrees of disorder (Martin et al. 2011). As the strength of disorder increases, Bloch eigenmodes begin to be localized, eventually exhibiting wavelength-scale Anderson localization (Fig. 3.3b–d). The localization leads to the suppression of wave transport inside the system (Fig. 3.3e–g).

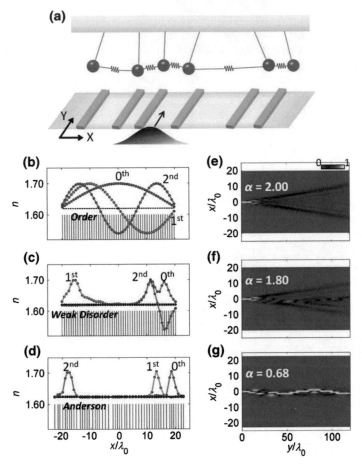

Fig. 3.3 Effect of disorder in photonics. **a** A schematic of a random optical system composed of coupled waveguides, analogous to randomly coupled pendulums with identical oscillating features. **b–d** The first three eigenmodes for **b** ordered, **c** weakly disordered, and **d** strongly disordered potentials. The potential n denotes the effective waveguide index of a single waveguide. Corresponding wave transports are shown in **e–g**, respectively. $\gamma_{00} = 1.6 \cdot k_0$, $\kappa_0 = 0.01 \cdot k_0$, and $N = 51$ for **b–g**, where $k_0 = 2\pi/\lambda_0$ is the free-space wavenumber. Figure and caption adapted from Yu et al. (2016a) under AAAS's License to Publish (http://www.sciencemag.org/help/reprints-and-permissions)

How can we quantify wave localization and transport? These behaviours can be quantified by two quantities, the modal size and diffusion exponent. First, the localization property is quantified by the modal size of each eigenmode. For example, we consider the 1D system of $\xi = y$, which has eigenmodes $\psi_k(x)$ and corresponding eigenvalues γ_k ($k = 1, 2, ..., N$ and γ_k is the effective wavevector of ψ_k). The modal size for each eigenmode is then defined as (Schwartz et al. 2007)

$$w_k = \frac{\left[\sum_{p=1}^{N} \left|\psi_k(x_p)\right|^2 \cdot \Delta x_p\right]^2}{\sum_{p=1}^{N} \left|\psi_k(x_p)\right|^4 \cdot \Delta x_p}, \quad (3.3)$$

where Δx_p is the pth element size obtained from the distance between waveguides.

The efficiency of wave transport is quantified by the diffusion exponent α (Hahn et al. 1996). For a 1D system ($\xi = y$) with an arbitrary incidence $\varphi_i(x) = \Sigma a_k \psi_k$, the field is $\varphi(x, y) = \Sigma a_k \psi_k exp(i\gamma_k y)$. When the incident wave is excited at the centre waveguide ($x = x_m$, where $m = (N + 1)/2$ for odd N), the spatially varying mean-square displacement (MSD) $M(y)$ is calculated as follows:

$$M(y) = \langle x^2 \rangle = \frac{\sum_p (x_p - x_m)^2 \cdot \left|\phi(x_p, y)\right|^2}{\sum_p \left|\phi(x_p, y)\right|^2}. \quad (3.4)$$

By fitting the MSD $M(y)$ for y as $M(y) \sim c_\alpha \cdot y^\alpha$, the diffusion exponent α is $\alpha = 2$ for ballistic transport and $\alpha = 1$ for diffusive transport (Hahn et al. 1996).

Figure 3.4 shows the quantification of wave characteristics using the modal size and diffusion exponent. With eigenmode localization (Fig. 3.4a) and the following forms of suppressed wave transport (Fig. 3.4b), increasing disorder also alters the spreading of the corresponding eigenvalues, linearizing the eigenband (Fig. 3.4c). A continuous transition between the regimes of order and disorder is then evident, confirming the classic relationship (Anderson 1958; Schwartz et al. 2007; Wiersma 2013; Wiersma et al. 1997; Lahini et al. 2008; Papasimakis et al. 2009; Poddubny et al. 2012) between localization, transport, and disorder.

Motivated by the decomposition of the Hamiltonian $H = D + K$, we demonstrate that the design of unconventional eigenmodes in disordered potentials can be

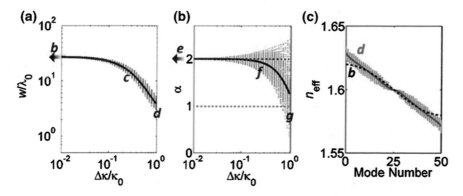

Fig. 3.4 Variations of **a** modal size w and **b** diffusion exponent α as a function of the disorder $\Delta\kappa$. **c** Eigenbands (n_{eff}) for disorder in Fig. 3.3b (black) and Fig. 3.3d (green). The points in (**a**–**c**) represent each ensemble, and the solid lines are the averages of 200 ensembles. The line of $\alpha = 1$ in (**b**) denotes the diffusion state. Figure and caption adapted from Yu et al. (2016a) under AAAS's License to Publish (http://www.sciencemag.org/help/reprints-and-permissions)

achieved by using the DOFs on D, the "on-site energy" of each element, which was neglected in Figs. 3.3 and 3.4.

Moulding of an Eigenmode The total DOFs in an N-body system are defined by an $N \times N$ Hamiltonian matrix H, which allows N^2 DOFs. On the other hand, the DOFs allowed in an already randomly coupled system with a given K are defined by the on-site energy of each element shown in the diagonal terms of D, leading to N DOFs. Thus, we can handle at most a "single" eigenmode among N eigenmodes in an N-body system, from N modes $\times$ ("N DOFs" of D/"N^2 DOFs" of H) = 1 mode.

Therefore, suppose that we desire to "mould" a single target eigenmode ψ to have the *spatial* distribution of $v_m = [v_{m1}, v_{m2}, ..., v_{mN}]^T$ with the eigenvalue γ_m in the randomly coupled system. For this purpose, we develop the inverse design technique in reciprocal space by assigning the spatial profile of v_m to one of the basis vectors. From this assignment, we introduce the eigen-decomposition matrix $V = [v_m, v_2, ..., v_N]$ using the Gram-Schmidt process, where v_m and the set of column vectors $v_s = [v_{s1}, v_{s2}, ..., v_{sN}]^T$ ($s = 2, 3, ..., N$) together compose the orthonormal basis. With the orthonormality ($VV^\dagger = I$), the equation in the V-reciprocal space becomes $V^\dagger HV(V^\dagger \psi) = \gamma(V^\dagger \psi)$, or $H_r \psi_r = \gamma \psi_r$, where $H_r = V^\dagger HV$ and $\psi_r = V^\dagger \psi$. The randomly coupled network K of the system is then represented by a complex matrix $K_r = V^\dagger KV$ in the V-reciprocal space. The V-reciprocal Hamiltonian $H_r = V^\dagger DV + K_r$ has the following components:

$$
H_r = \begin{bmatrix}
\sum_p v_{mp}^* \gamma_{op} v_{mp} + K_{r11} & \sum_p v_{mp}^* \gamma_{op} v_{2p} + K_{r12} & \cdots & \sum_p v_{mp}^* \gamma_{op} v_{Np} + K_{r1N} \\
\sum_p v_{2p}^* \gamma_{op} v_{mp} + K_{r21} & \sum_p v_{2p}^* \gamma_{op} v_{2p} + K_{r22} & \cdots & \sum_p v_{2p}^* \gamma_{op} v_{Np} + K_{r2N} \\
\vdots & \vdots & \ddots & \vdots \\
\sum_p v_{Np}^* \gamma_{op} v_{mp} + K_{rN1} & \sum_p v_{Np}^* \gamma_{op} v_{2p} + K_{rN2} & \cdots & \sum_p v_{Np}^* \gamma_{op} v_{Np} + K_{rNN}
\end{bmatrix},
$$

$$(3.5)$$

where $\gamma_{op} = [D]_p$ and $K_{rpq} = [K_r]_{pq}$. The relationship $H_r = V^\dagger HV$, which is equivalent to the intertwining relationship of $HV = VH_r$ for the orthonormal vector V ($VV^\dagger = I$), represents the isospectrality between H and H_r, as shown in Sect. 2.1.

To design the eigenmode ψ of spatial representation v_m, we set one of the reciprocal eigenmodes to be $\psi_r = [1, 0, ..., 0]^T$. This condition is uniquely fulfilled when the 1st column of H_r has only one nonzero component of $[H_r]_{11}$. From the DOF of this nonzero component, the target eigenvalue can be set by assigning $[H_r]_{11} = \gamma_m$. From $[[H_r]_{11}, [H_r]_{21}, ..., [H_r]_{N1}]^T = [\gamma_m, 0, ..., 0]^T$ with the expression of Eq. (3.5), we achieve the on-site energy of each element γ_{op} deterministically as:

$$
\begin{bmatrix}
\gamma_{o1} \\
\gamma_{o2} \\
\vdots \\
\gamma_{oN}
\end{bmatrix}
=
\left(
\begin{bmatrix}
v_{m1}^* & v_{m2}^* & \cdots & v_{mN}^* \\
v_{21}^* & v_{22}^* & \cdots & v_{2N}^* \\
\vdots & \vdots & \ddots & \vdots \\
v_{N1}^* & v_{N2}^* & \cdots & v_{NN}^*
\end{bmatrix}
\begin{bmatrix}
v_{m1} & 0 & \cdots & 0 \\
0 & v_{m2} & \cdots & 0 \\
\vdots & \vdots & \ddots & \vdots \\
0 & 0 & \cdots & v_{mN}
\end{bmatrix}
\right)^{-1}
\begin{bmatrix}
\gamma_m - K_{r11} \\
-K_{r21} \\
\vdots \\
-K_{rN1}
\end{bmatrix}, \quad (3.6)
$$

or, simply, $\gamma_o = [\gamma_{o1}, \gamma_{o2}, ..., \gamma_{oN}]^T = [diag(v_m)]^{-1}V[\gamma_m - K_{r11}, -K_{r12}, ..., -K_{r1N}]^T$.

Equation (3.6) indicates that if the inverse of the matrix $diag(v_m)$ exists, i.e., $v_{mi} \neq 0$ for all i, the on-site energy vector γ_o can always be found. Because the on-site energy vector represents the distribution of optical potentials, such as the size or refractive index of each waveguide that determines its wavevector, we prove that the nodeless eigenmode ($v_{mi} \neq 0$) uniquely determines the optical potential of each element in weakly coupled systems. A situation similar to the 1:1 correspondence between a nodeless eigenmode and a wave potential can be found in quantum mechanics, the relationship between the ground state and the necessary quantum mechanical potential in the Schrödinger equation. However, the main difference between these scenarios lies in the presence of the random network matrix K in weakly coupled systems. Because K includes the information of the spatial distribution of each element, the on-site energy vector γ_o does not represent the entire optical potential profile in real space.

To summarize, for "*any*" networks K regardless of the degree of disorder, we show that at least a "single" eigenmode can always be moulded into the desired shape with the target eigenvalue by adjusting the potential of each element. The proposed system is included in the concept of metadisorder, the engineered D with the random K. The proposed metadisorder system allows the unconventional form of *an eigenmode* in all regimes of random networks K, for example, the globally collective eigenmode. **The process of moulding an eigenmode to manipulate the properties of metadisorder structures** is as follows (**P- for process**):

P-I. We have a given random matrix K that represents the spatial distribution of each optical element.

P-II. We set the spatial distribution of a target eigenmode.

P-III. From Eq. (3.6), **we derive the on-site energy distribution (matrix D)**, and the following Hamiltonian $H = D + K$, which includes the complete information of the system.

P-IV. We examine the designed coupled system for wave phenomena related to disordered photonics.

As an example, we begin with 1D metadisorder structures.

1D Metadisorder Figure 3.5 shows examples of designer eigenmodes in 1D metadisorder systems, where the optical potential of each waveguide is calculated via Eq. (3.6). Compared with the conventional off-diagonal disorder in Fig. 3.3, the metadisorder platform allows the construction of various systems with a designer eigenmode v_m including a plane wave (Fig. 3.5b), Gaussian-enveloped guided wave with nonexponential spatial decay (Fig. 3.5c), and both interface (Fig. 3.5d) and surface (Fig. 3.5e) waves with Anderson-type exponential decay (Anderson 1958; Schwartz et al. 2007; Martin et al. 2011; Lahini et al. 2008). The real-valued v_m corresponds to the ground state in the disordered eigenband (red arrows in Fig. 3.5f–i).

In detail, our design method allows a scattering-free plane wave despite the disordered coupling network inside the system (Fig. 3.5b, $\Delta\kappa = \kappa_0$), in stark contrast to conventional disorder (Fig. 3.3g), which leads to strong localization from random scattering. Interestingly, the plane wave eigenmode in the metadisorder system,

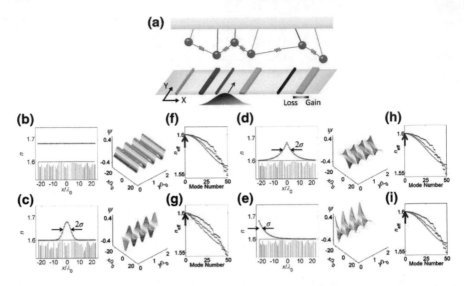

Fig. 3.5 1D metadisorder systems. a A schematic of a 1D metadisorder system composed of coupled waveguides, analogous to the randomly coupled pendulums with different self-oscillating features, such as an oscillating period (rod length) and gain or loss parameters (colour). Each waveguide has different real parts of self-energy due to changing the width of the waveguide. The colours of the waveguides represent the imaginary parts of the self-energy: gain and loss. **b–e** Designed eigenmodes and optical potentials, eigenmode propagations, and **f–i** eigenvalues (n_{eff}) of 1D metadisorder systems are calculated for **b, f** plane wave ($v_m(x) = 1$), **c, g** Gaussian wave ($v_m(x) = exp[-x^2/(2 \cdot \sigma^2)]$), **d, h** interface wave ($v_m(x) = exp[-|x|/(2 \cdot \sigma)]$), and **e, i** surface wave ($v_m(x) = exp[-|x - x_L|/(2 \cdot \sigma)]$) eigenmodes, where the spatial bandwidth $\sigma = L_{st}/16$ in **c–e**, the left boundary $x_L = -L_{st}/2$, and L_{st} is the overall potential length. $\Delta\kappa = \kappa_0$ in **b–e** for the extreme degree of disorder. Blue symbols represent $\Delta\kappa = \kappa_0$, green symbols represent $\Delta\kappa = 0.53 \cdot \kappa_0$ and black dotted lines represent $\Delta\kappa = 0$ in **f–i**. Figure and caption adapted from Yu et al. (2016a) under AAAS's License to Publish (http://www.sciencemag.org/help/reprints-and-permissions)

which has a modal size equal to the overall system size, can be more extended than that in finite-N ordered systems with field discontinuity at the boundary due to broken translational symmetry (Fig. 3.3b). Therefore, the designer modulation of the on-site energy distribution allows the cancellation of the scattering not only from the random coupling network inside the system but also from the boundary discontinuity. Furthermore, we can obtain unconventional forms of wave localization, such as non-Anderson Gaussian localization (Fig. 3.5c) or designed Anderson-type exponential localization at the interface (Fig. 3.5d) or surface (Fig. 3.5e), as a generalization of the accidental emergence of classic Anderson localization (Fig. 3.3d).

A potential with a globally extended eigenmode should have reduced random scattering in the overall system, especially in terms of the phase information, maintaining the coherence between multiple scatterings. Therefore, other eigenmodes of "similar" eigenvalues, which undergo multiple scatterings of a similar strength, also tend to have wider spatial bandwidth than those of conventional disorder. This phenomenon allows an overall increase in the modal size in the set of eigenmodes, which has been generalized to the concept of non-Anderson disorder, the realization of modal delocalization at the same level of structural disorder (Yu et al. 2018a).

"Small-World" in Metadisorder We have explored the property of individual eigenmodes and eigenvalues inside metadisorder structures. We need to also investigate optical phenomena that arise from the collective contributions of several eigenmodes, including scattering and wave transport. The concept of metadisorder creates a new class of wave phenomena in these phenomena that derive a counterintuitive relationship between the eigenmode localization (w) and wave transport (α) similar to the small-world phenomena in graph networks.

In Fig. 3.6, we consider the localized designer eigenmode of the specific form $v_m(x) = exp[-\ |x|^g/(2 \cdot \sigma^g)]$ (Fig. 3.6a), where $g = 1$ for Anderson-type exponential localization and $g = 2, 4$, and 6 for the examples of non-Anderson localizations. Figure 3.6f–n presents the localization-transport (w–α) relationship of 1D non-Anderson metadisorder systems compared with conventional Anderson disorder (Fig. 3.6b) and Anderson-type metadisorder (Fig. 3.6c–e). Although Anderson-type metadisorder ($g = 1$) provides similar w–α relationships to that of Anderson disorder (Fig. 3.6c–e vs. Fig. 3.6b), non-Anderson metadisorder ($g > 1$) enables more "localized" waves yet achieves ballistic transport (e.g., Fig. 3.6f vs. Fig. 3.6b, a factor of ~2 decrease for w and α ~ 2 for $\Delta\kappa < \kappa_0/10$). This opposite and separate con-

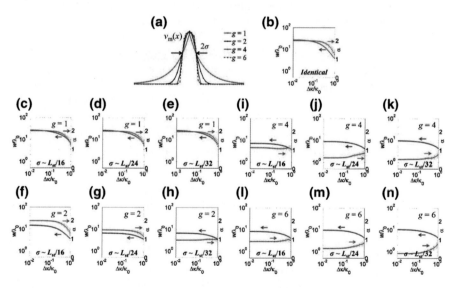

Fig. 3.6 Small-world-type metadisorder systems with disorder-induced wave transport. **a** Shapes of designer eigenmodes $v_m(x) = exp[-\ |x|^g/(2 \cdot \sigma^g)]$ with different g. Eigenmode localization (w) and wave transport (α) for **b** Anderson disorder with identical elements, **c–e** $g = 1$ Anderson-type metadisorder, and **f–h** $g = 2$, **i–k** $g = 4$, and **l–n** $g = 6$ non-Anderson metadisorder. In metadisorder systems, the bandwidths of the designer eigenmodes are $\sigma = L_{st}/16$ in **c, f, i, l**, $\sigma = L_{st}/24$ in **d, g, j, m**, and $\sigma = L_{st}/32$ in **e, h, k, n**. Error bars in **b–n** denote the standard deviation of 200 ensembles. Figure and caption adapted from Yu et al. (2016a) under AAAS's License to Publish (http://www.sciencemag.org/help/reprints-and-permissions)

trol of modal localization and wave transport is analogous to the separate control of localization and transport in "small-world" networks (Watts and Strogatz 1998b). Such an anomalous wave transport even enables "disorder-induced" wave transport (Fig. 3.6h), showing the reverse relationship between w and α compared with that of conventional disorder (Schwartz et al. 2007; Garanovich et al. 2012; Wiersma 2013; Wiersma et al. 1997; Lahini et al. 2008). This counterintuitive relationship is more apparent for metadisorder with larger g (Fig. 3.6i–n for $g = 4$ and 6), allowing not only the separate control of localization and transport with g and σ but also the robustness of wave transport to the disorder $\Delta \kappa$, analogous to the difference between clustering and path length in small-world networks (Watts and Strogatz 1998b).

3.3 Application I: Functional Wave Transport

Although properties such as small-world-like behaviours are interesting, what are the "obvious" benefits of using the metadisorder instead of ordered structures? Here, we show examples of applications of metadisorder with an extension to a higher dimension and discuss the advantages of metadisorder.

2D Hermitian Metadisorder The reciprocal space design allows for an extension to multidimensional problems by including all the existing coupling coefficients in the network matrix K. We consider 2-dimensional (2D) metadisorder structures designed with a randomly deformed lattice (Fig. 3.7a). At this stage, structures are defined in the regime of Hermitian Hamiltonians with real-valued potentials.

Figure 3.7b represents an example of deformed 17×17 square lattices. Figure 3.7c–h shows collective wave behaviours in 2D metadisorder systems, which support strong coherence over the system: free-form standing-wave resonances with a uniform field distribution (Fig. 3.7c), quadrupole phase distribution (Fig. 3.7d), and the designed localization (Fig. 3.7e) despite the random deformation of the coupling network.

2D Non-Hermitian Metadisorder The inverse design technique based on the moulding of an eigenmode in Eq. (3.6) is independent from the Hermiticity of the system Hamiltonian. By setting the spatial profile of the target eigenmode properly, we can achieve metadisorder structures in the non-Hermitian regime (Bender and Boettcher 1998; Feng et al. 2017; El-Ganainy et al. 2018), which include the elements composed of gain or loss materials.

For example, consider the target eigenmode for "one-way" traveling-wave resonances by imposing the form $exp(-i\rho\theta)$ on v_m (Fig. 3.8a–c for the azimuth θ). Such a phase evolution inside the eigenmode is one of the important characteristics in parity-time (PT) symmetric (or, more broadly, non-Hermitian) potentials (Guo et al. 2009; Longhi 2009a; Rüter et al. 2010; Regensburger et al. 2012; Yu et al. 2012, 2013, 2015a, b, 2017a, 2018b; Feng et al. 2013; Ge and Stone 2014; Hodaei et al. 2014; Peng et al. 2014; Assawaworrarit et al. 2017; Teimourpour et al. 2016, 2017a,

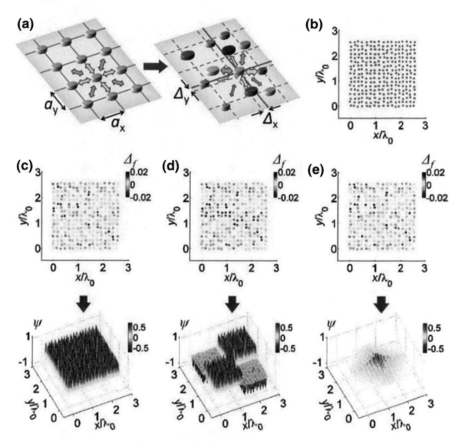

Fig. 3.7 2D Hermitian metadisorder systems. **a** A lattice with identical elements (left) and its metadisorder counterpart with the designed on-site energy of each element (right). **b** A sample of the deformed 17×17 lattice. **c–e** 2D metadisorder systems: **c–e** collective resonances for **c** uniform, **d** quadrupole, and **e** exponentially localized distributions. The designed resonance is set to $\gamma_m = f_0$ for all cases. Figure and caption adapted from Yu et al. (2016a) under AAAS's License to Publish (http://www.sciencemag.org/help/reprints-and-permissions)

b; Ge and El-Ganainy 2016; Yu et al. 2019b). This feature has been intensively studied in polarization optics, the emergence "chiral" eigenpolarization defined by the basis in PT-symmetric Hamiltonians (Lawrence et al. 2014; Yu et al. 2016b, c, 2018c, 2019a; Piao et al. 2018; Yu et al. 2019a). In a 2D in-plane metadisorder structure, the "chiral" rotation of the phase in the collective wave evolutions derives (i) the travelling wave resonances in coupled resonator structures and (ii) the orbital angular momentum (OAM) (Molina-Terriza et al. 2007) in coupled waveguide structures. Notably, although the chiral features of light inside the proposed metadisorder systems requires complex optical potentials with gain and loss materials, our approach involves neither PT symmetry nor periodicity.

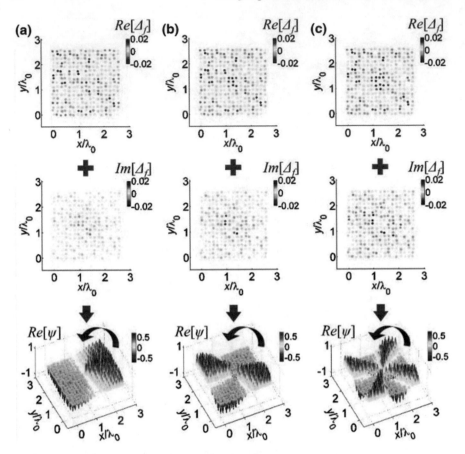

Fig. 3.8 2D non-Hermitian metadisorder systems. **a–c** Traveling-wave chiral collective resonances using complex potentials for **a** dipole, **b** quadrupole, and **c** octupole modes. Figure and caption adapted from Yu et al. (2016a) under AAAS's License to Publish (http://www.sciencemag.org/help/reprints-and-permissions)

2D Functional Metadisorder Having demonstrated collective modes in highly disordered systems with both Hermitian and non-Hermitian Hamiltonians, we now present the excitation technique of designer eigenmodes with the external coupling of conventional waveforms. When the inner connection of the metadisorder system is sufficiently strong, which allows the viewpoint based on a photonic molecule, the separation of eigenvalues that determines the free spectral range (FSR) becomes large enough to achieve wave phenomena solely dependent on a target eigenmode.

First, Fig. 3.9a shows the wave flow through the perfectly uniform collective eigenmode over the entire metadisorder system. Following the property of the designed eigenmode, the details of disordered profiles in the system become "invisible" for incident plane waves, suppressing any alteration of phase and amplitude (transmission $T \sim 100\%$) and thus realizing zero effective index. With eigenmode-based

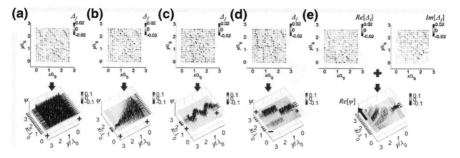

Fig. 3.9 2D functional metadisorder systems. **a–e** Metadisorder-based functionalities in the 17 × 17 lattice for **a** invisible disorder, **b** steered focusing, **c** spatial oscillation, **d** parity-converted beam splitter, and **e** point-source excitation of oblique plane waves. The fields in the input and output are multiplied by 40. Figure and caption adapted from Yu et al. (2016a) under AAAS's License to Publish (http://www.sciencemag.org/help/reprints-and-permissions)

metadisorder design, we also implement high-level photonic functionalities with excellent throughputs, including tunable focusing of light waves (Fig. 3.9b, $T \sim 96\%$), phase-conserved oscillation (Fig. 3.9c, $T \sim 98\%$), parity mode converters (even to odd, Fig. 3.9d, $T \sim 99\%$) of real potentials, and excitation of oblique plane waves with a point-source ($5.6°$, Fig. 3.9e, $T \sim 97\%$) using complex potentials.

Metadisorder Robustness The above functionalities can also be achieved in ordered systems, such as periodic systems with proper structural defects. Then, what is the advantage of metadisorder structures? In Fig. 3.10, we test the stability of the delocalization against on-site energy and coupling errors, taking an example of plane wave metadisorder structures. We compare a randomly coupled metadisorder system that leads to a plane wave eigenmode (Fig. 3.5b) with a periodic system in which the components possess identical on-site energy (Fig. 3.3b). For both systems, we apply white-noise error both to the on-site energy (Fig. 3.10a, b, $\rho_\gamma \cdot u(0, 1)$) and coupling coefficient (Fig. 3.10c, d, $\rho_\kappa \cdot u(0, 1)$), where ρ_γ and ρ_κ are the relative magnitudes of the white noise to the original terms γ_{00} and κ_0, respectively. These errors represent the imperfections in the (i) size or refractive index of each element (waveguides or resonators) and in the (ii) distance between elements during the fabrication process.

In terms of the error robustness, the metadisorder structure supports a level of tolerance similar to that of the ideal periodic structure for both on-site energy and coupling errors. Notably, among the statistical ensemble of 400 random samples, a set of metadisorder systems exists that provides even more robust function (here, delocalization) than that of the periodic systems against the errors. This result demonstrates that the metadisorder system provides stable delocalization in randomly coupled systems.

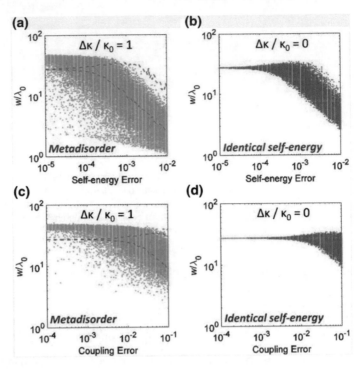

Fig. 3.10 Robustness of the metadisorder system, estimated by the variation of the ground state modal size against the **a, b** on-site energy errors and **c, d** coupling errors, for **a, c** the metadisorder system ($\Delta\kappa/\kappa_0 = 1$) and **b, d** the periodic system ($\Delta\kappa/\kappa_0 = 0$) with identical elements. Figure and caption adapted from Yu et al. (2016a) under AAAS's License to Publish (http://www.sciencemag.org/help/reprints-and-permissions)

3.4 Application II: Cloaking Inside Coupled Systems

In Sects. 3.2 and 3.3, we assume that the target eigenmode is nodeless, i.e., $v_{mi} \neq 0$ for all i, for the use of Eq. (3.6). What happens when nodes exist inside the target eigenmode? At a glance, since the optical potential at the node of the target eigenmode does not affect the characteristics of that mode, we can achieve the DOFs to handle the other properties of the light wave. From this viewpoint, we introduce the target decoupling technique inside photonic coupled systems, the cloaking of the target elements inside coupled networks.

Why do we need Target Decoupling? Coupling enables quantized behaviours of light (Christodoulides et al. 2003) that are impossible in the bulk. Thus, photonic coupled systems serve as building blocks for photonic signal processing in integrated platforms (Takesue et al. 2013; Song et al. 2015) and classical analogy of quantum mechanics (Yu et al. 2015c; Aspuru-Guzik and Walther 2012) based on the tunable selectivity of spectral and momentum domains. However, inevitable coupling between nearby elements could also cause unwanted crosstalk, which has been the major hurdle for designer wave flows in integrated systems (Mrejen et al. 2015a; Nagarajan et al. 2005). Thus, the realization of decoupling techniques for densely packed coupled systems is an urgent issue.

The decoupling of optical elements can be reinterpreted in terms of "hiding" optical elements inside coupled systems. In this context, invisibility cloaking is one of the most fascinating achievements in metamaterials and transformation optics (Pendry et al. 2006; Chen et al. 2010; Xu and Chen 2015). Although transformation optics derived from full-vectorial Maxwell's equations successfully provides an exact solution for omnidirectional and scattering-free perfect cloaking, at the same time, its strict demand on material designs has caused hardship to the application of the cloaking in coupled optical systems. For example, consider the hiding of some elements inside coupled optical systems (Takesue et al. 2013; Yu et al. 2015c; Shafiei et al. 2013; Hsieh et al. 2015; Mrejen et al. 2015b). The conventional transformation optics (Pendry et al. 2006; Chen et al. 2010; Xu and Chen 2015) then provide severely intricate solutions even for the approximated case (Cai et al. 2007): the coating of target elements with highly anisotropic metamaterials (Cai et al. 2007) of extreme material parameters ($\varepsilon \sim 0$), which enables the "microscopic" removal of the coupling to the target elements.

Similar restrictions also exist in other alternative cloaking techniques. The cloaking using accidental degeneracy for effective zero index (Huang et al. 2011) requires the well-defined and very sensitive crystalline structure to maintain the Dirac point. Therefore, cloaked elements, which will be defined as defects inside photonic crystals, should be separated by more than several lattice periods, prohibiting the integration. Although the concept of PT symmetry has been applied to the unidirectional invisibility in one-dimensional coupled systems (Longhi 2011; Lin et al. 2011) using nonorthogonal eigenmodes and the following singular scattering, the extension to multi-dimensional integrated systems encounters the similar difficulty with transformation optics: the coating of spatially varying gain or loss media (Zhu et al. 2013) for each element with rigorous symmetry. The optical analogy of the adiabatic passage (Mrejen et al. 2015b; Longhi et al. 2007) has also been employed to achieve the decoupling in tri-atomic designs, but its multi-dimensional or N-atomic realization still remains as a challenge.

As a solution to this problem, we adopt the "macroscopic" approach to the decoupling which is applicable to coupled optical systems. In contrast to conventional cloaking techniques with "microscopic" material arrangement for each element, we use the designer system eigenmode encompassing decoupled elements, providing excellent flexibility to the waveform moulding in coupled optical systems.

Concept of Target Decoupling We begin with an instructive example: a triatomic system where each element has the on-site energy of ρ_i, and the coupling between the ith and jth elements is given as κ_{ij} (Fig. 3.11a, $\kappa_{ij} \sim \kappa_{ji}$ for the similar shape of elements (Haus 1984)). The system satisfies the Hamiltonian equation (Christodoulides et al. 2003; Mrejen et al. 2015b):

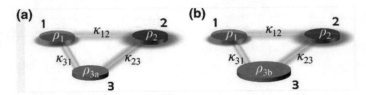

Fig. 3.11 Target decoupling in tri-atomic examples for the different 3rd element: **a** ρ_{3a} and **b** ρ_{3b}. Figure and caption adapted from Yu et al. (2017b) under a CC BY license (http://creativecommons. org/licenses/by/4.0/)

$$
\begin{bmatrix}
\rho_1 & \kappa_{12} & \kappa_{13} \\
\kappa_{21} & \rho_2 & \kappa_{23} \\
\kappa_{31} & \kappa_{32} & \rho_3
\end{bmatrix}
\begin{bmatrix}
\psi_1 \\
\psi_2 \\
\psi_3
\end{bmatrix}
= \rho
\begin{bmatrix}
\psi_1 \\
\psi_2 \\
\psi_3
\end{bmatrix},
\tag{3.7}
$$

for the field amplitude of each element $\Psi = [\psi_1, \psi_2, \psi_3]^{\mathrm{T}}$. We try to achieve the decoupling of the 3rd element, which requires the eigenmode independent from the random perturbation of ρ_3 (Fig. 3.11a vs. Fig. 3.11b, as $\rho_{3a} \neq \rho_{3b}$). This condition is satisfied with $\psi_3 = 0$, which corresponds to the "hiding" of the 3rd element for the target eigenmode. Equation (3.7) then derives $\kappa_{31} \cdot \psi_1 + \kappa_{32} \cdot \psi_2 = 0$, forming the destructive coupling interference in the 3rd element. This condition again applied to Eq. (3.7) then derives the necessary condition of the on-site energy for decoupling the 3rd element, as: $\rho_1 - \rho_2 = \kappa_{31} \cdot \kappa_{12}/\kappa_{32} - \kappa_{32} \cdot \kappa_{21}/\kappa_{31}$. The corresponding eigenvalue of the target eigenmode can be controlled by $\rho = \rho_1 - \kappa_{31} \cdot \kappa_{12}/\kappa_{32}$. Therefore, by controlling the on-site energy of the elements ($\rho_{1,2}$) with the given coupling network (fixed κ_{ij}), we can "hide" some elements inside the coupled system at the desired eigenvalue ρ. In this case, the coupling network can be any types of networks, even including irregular or symmetry-broken cases (e.g., $\kappa_{23} \neq \kappa_{31}$).

Target Decoupling in N-atomic Systems The suggested approach is easily extended to hiding m-elements inside N-atomic systems. The key of design criteria is the construction of "shielding" for each decoupled element based on destructive interferences. First, the equation for N-atomic systems composed of weakly coupled elements is (Christodoulides et al. 2003; Yu et al. 2016a; Mrejen et al. 2015b; Longhi 2009b)

$$
\begin{bmatrix}
\rho_1 & \kappa_{12} & \cdots & \kappa_{1N} \\
\kappa_{21} & \rho_2 & \cdots & \kappa_{2N} \\
\vdots & \vdots & \ddots & \vdots \\
\kappa_{N1} & \kappa_{N2} & \cdots & \rho_N
\end{bmatrix}
\begin{bmatrix}
\psi_1 \\
\psi_2 \\
\vdots \\
\psi_N
\end{bmatrix}
= \rho \cdot
\begin{bmatrix}
\psi_1 \\
\psi_2 \\
\vdots \\
\psi_N
\end{bmatrix}.
\tag{3.8}
$$

We then derive the necessary form of the decoupling eigenmode $\Psi = [\psi_1, \psi_2, \ldots, \psi_N]^{\mathrm{T}}$ for the given network (κ_{jk}, Fig. 3.12a), to determine the on-site energy (Yu et al. 2016a) $\Omega = [\rho_1, \rho_2, \ldots, \rho_N]^{\mathrm{T}}$.

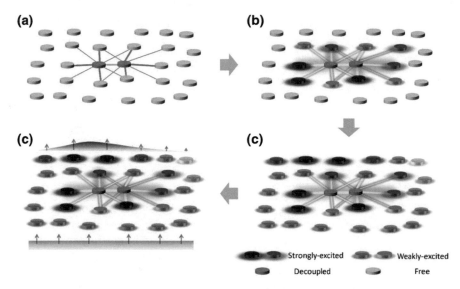

Fig. 3.12 The design procedure of an N-atomic system for a decoupling eigenmode. **a** The selection of target decoupled elements in the system which has the determined coupling network. The design of the field distribution in **b** nearby elements of target elements and **c** the rest tunable elements. **d** The wave flow through the system, realizing the scattering-free manipulation of the waveform. Couplings only around the decoupled elements are presented for clarity. Figure and caption adapted from Yu et al. (2017b) under a CC BY license (http://creativecommons.org/licenses/by/4.0/)

With m number of decoupled elements ($m \leq N$, $\psi = 0$) the indices of which constitute the set A, the set of nearby elements ($\psi \neq 0$) for each decoupled element can be defined as B_j ($j \in$ A). Because at most nearest-neighbour and next-nearest-neighbour coupling coefficients are significant in realistic structures (Keil et al. 2015) (Fig. 3.12a) due to the exponential decay of evanescent coupling in space, each row of Eq. (3.8) for decoupled elements derives the following condition of the destructive coupling interference as

$$\sum_{k \in B_j} \kappa_{jk} \cdot \psi_k = 0, \tag{3.9}$$

where $j \in$ A, and the condition of $k \in B_j$ represents the nearby coupling ($\kappa_{jk} \sim 0$ for far-off elements of $k \notin B_j$, for the j-th decoupled element). Equation (3.9) governs the necessary condition of the field amplitude in nearby elements (Fig. 3.12b), providing each equation for $j \in$ A which has the number of variables equal to the number of elements in B_j. The DOF of Eq. (3.9), the difference between the number of variables (field amplitude in nearby elements) and the number of constraints by equations (zero-field in decoupling elements), is determined by the number of nearby elements for each decoupled element. After the nearby elements for all of the decoupled elements are designed to satisfy Eq. (3.9) (Fig. 3.12b, the elements in B_j for all $j \in$ A), the other region of the decoupling eigenmode can then be designed (Fig. 3.12c). Except the rows of decoupled indices for Eq. (3.9), the other part of Eq. (3.8) has the form of

$$
\begin{bmatrix}
\rho_{s-1} & \kappa_{s-1,s-2} & \cdots & \kappa_{s-1,s-(N-m)} \\
\kappa_{s-2,s-1} & \rho_{s-2} & \cdots & \kappa_{s-2,s-(N-m)} \\
\vdots & \vdots & \ddots & \vdots \\
\kappa_{s-(N-m),s-1} & \kappa_{s-(N-m),s-2} & \cdots & \rho_{s-(N-m)}
\end{bmatrix}
\begin{bmatrix}
\psi_{s-1} \\
\psi_{s-2} \\
\vdots \\
\psi_{s-(N-m)}
\end{bmatrix}
= \rho \cdot
\begin{bmatrix}
\psi_{s-1} \\
\psi_{s-2} \\
\vdots \\
\psi_{s-(N-m)}
\end{bmatrix},
$$

$$(3.10)$$

where the new index $(s - p) \notin A$ and $1 \leq (s - p) \leq N$. For the subset of the decoupling eigenmode $\Psi_s = [\psi_{s-1}, \psi_{s-2}, \ldots, \psi_{s-(N-m)}]^T$, although the nearby elements of decoupled elements $((s - p) \in B_j$ for all $j \in A$, red and blue elements in Fig. 3.12b) are already determined for the decoupling (Eq. (3.9)), the other elements $((s - p) \notin B_j$ for any $j \in A$, light grey elements in Fig. 3.12c) can be freely set to achieve desired optical functionalities (e.g., steered beam focusing in Fig. 3.12d), defining Ψ_s and then Ψ where $\psi_j = 0$ for $j \in A$.

From the decoupling eigenmode Ψ with the desired function, we derive the corresponding on-site energy distribution $\Omega = [\rho_1, \rho_2, \ldots, \rho_N]^T$. The $(N - m) \times (N - m)$ matrix equation of Eq. (3.10), off-diagonal terms of which have given values, can be recast into the form of

$$
\begin{bmatrix}
\psi_{s-1} & 0 & \cdots & 0 \\
0 & \psi_{s-2} & \cdots & 0 \\
\vdots & \vdots & \ddots & \vdots \\
0 & 0 & \cdots & \psi_{s-(N-m)}
\end{bmatrix}
\begin{bmatrix}
\rho_{s-1} \\
\rho_{s-2} \\
\vdots \\
\rho_{s-(N-m)}
\end{bmatrix}
$$

$$
= \left(\rho \cdot I -
\begin{bmatrix}
0 & \kappa_{s-1,s-2} & \cdots & \kappa_{s-1,s-(N-m)} \\
\kappa_{s-2,s-1} & 0 & \cdots & \kappa_{s-2,s-(N-m)} \\
\vdots & \vdots & \ddots & \vdots \\
\kappa_{s-(N-m),s-1} & \kappa_{s-(N-m),s-2} & \cdots & 0
\end{bmatrix}
\right)
\begin{bmatrix}
\psi_{s-1} \\
\psi_{s-2} \\
\vdots \\
\psi_{s-(N-m)}
\end{bmatrix},
$$

$$(3.11)$$

where I is the $(N - m) \times (N - m)$ identity matrix. Because the diagonal matrix $diag(\Psi_s)$ has its inverse due to $\psi_{(s-p)} \neq 0$ for all $(s - p) \notin A$, Eq. (3.11) derives the required on-site energy distribution $\Omega_s = [\rho_{s-1}, \rho_{s-2}, \ldots, \rho_{s-(N-m)}]^T$ except the decoupled elements as,

$$
\begin{bmatrix}
\rho_{s-1} \\
\rho_{s-2} \\
\vdots \\
\rho_{s-(N-m)}
\end{bmatrix}
=
\begin{bmatrix}
\psi_{s-1} & 0 & \cdots & 0 \\
0 & \psi_{s-2} & \cdots & 0 \\
\vdots & \vdots & \ddots & \vdots \\
0 & 0 & \cdots & \psi_{s-(N-m)}
\end{bmatrix}^{-1}
$$

$$\left(\rho \cdot I - \begin{bmatrix} 0 & \kappa_{s-1,s-2} & \cdots & \kappa_{s-1,s-(N-m)} \\ \kappa_{s-2,s-1} & 0 & \cdots & \kappa_{s-2,s-(N-m)} \\ \vdots & \vdots & \ddots & \vdots \\ \kappa_{s-(N-m),s-1} & \kappa_{s-(N-m),s-2} & \cdots & 0 \end{bmatrix} \right) \begin{bmatrix} \psi_{s-1} \\ \psi_{s-2} \\ \vdots \\ \psi_{s-(N-m)} \end{bmatrix}$$

$$(3.12)$$

Because the satisfaction of Eqs. (3.9) and (3.10) corresponds to the satisfaction of Eq. (3.8), the on-site energy distribution Ω which has the subset of Ω_s from Eq. (3.12) and "arbitrary" values for the Ω_s's complementary set, derives the decoupling eigenmode Ψ which has $\psi_j = 0$ for $j \in A$ and Ψ_s for the other part. The eigenmode Ψ in the potential Ω therefore achieves the decoupling (scattering-free for arbitrary ρ_j of $j \in A$) and the functionality (designed Ψ_s, Fig. 3.12d) at the same time.

Target Decoupling in Coupled Systems Based on the design methodology, we demonstrate the decoupling in coupled optical systems. We investigate the 11×11 coupled resonator square lattice, encompassing the 3×3 decoupled region at the centre of the system (the "decoupled" region D in Fig. 3.13. Its surrounding 'transport' region is denoted as T). The binary random on-site energy is applied to the resonators in the region D. By following the target decoupling methodology, the on-site energy distribution of the region T is derived both for the decoupling of the region D, and for the designed spatial profile of wave transport.

Figure 3.13 shows planewave spatial profiles, demonstrating the decoupling wave transfer for the sets of elements inside the target region D. Usually, the detailed configuration of the on-site energy distribution strongly affects the wave transport in the coupled optical system, because the on-site energy determines not only the phase evolution inside each element but also the coupling efficiency between elements

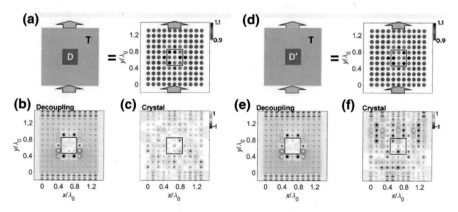

Fig. 3.13 Demonstration of eigenmode decoupling for planewave input and output waves through the crystal lattice. The different configurations in the decoupled region are compared for the cases of **a–c** and **d–f** (D $\neq$ D', red boxes in the right panels of **a**, **d**). The decoupling results in **b**, **e** are compared with the results of ordinary crystal systems in **c**, **f** composed of identical elements. λ_0 is the free-space wavelength. Figure and caption adapted from Yu et al. (2017b) under a CC BY license (http://creativecommons.org/licenses/by/4.0/)

(Haus 1984). However, regardless of the configuration of the target region D (D ≠ D′ in Fig. 3.13a, d), the eigenmode decoupling systems provide the perfect planewave transfer (Fig. 3.13b, e) with the same transport region T configuration, in sharp contrast to strong scattering and spatial incoherence in the crystal platforms the light flow of which has also strong dependence on the configuration of the region D (D ≠ D′ in Fig. 3.13c, f). This result demonstrates that the decoupling eigenmode successfully neglects the on-site energy perturbation inside the target region, realizing the "target decoupling" based on the form of the eigenmode.

Target Decoupling with Functionalities As shown in Eq. (3.12), the on-site energy distribution is uniquely defined for "any" nodeless eigenmode which satisfies the decoupling ($\psi = 0$) in the region D. Conversely, by controlling the on-site energy of the environment T (T′ in Fig. 3.14a), the moulding of wave flows becomes possible while preserving the scattering-free condition around D: the wave focusing in Fig. 3.14b (compared to the ordinary environment of Fig. 3.14c). Designer wave flows with functionalities, such as focusing or beam splitting can thus be achieved, regardless of the perturbation inside D.

The main strength of the eigenmode decoupling is the high applicability to "any" coupling networks, in contrast to the indispensable spatial symmetry in the Dirac point cloaking (Huang et al. 2011) or PT-symmetric invisibility (Longhi 2011; Lin et al. 2011; Zhu et al. 2013). The evidence is shown in Fig. 3.14d–f, demonstrating the decoupling in the system which has the off-diagonal disorder (Martin et al. 2011; Pendry 1982) from the random deformation of each resonator position (disordered coupling both in D_d and T_d regions in Fig. 3.14d). Perfect coherent transmission (Fig. 3.14e) is achieved as same as the cases in the lattice structure, overcoming the

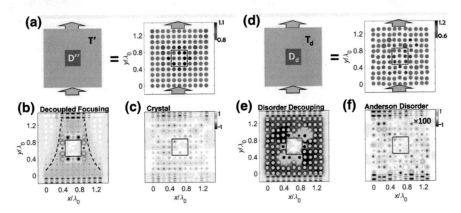

Fig. 3.14 Decoupling with functionalities of focusing and disorder-resistant transport. **a–c** The decoupling with wave focusing (T′): **a** a schematic, **b** the field profile in the decoupling system, and **c** the field profile in the ordinary crystal system. **d–f** The decoupling in the disordered system (D_d, T_d): **d** a schematic, **e** the field profile in the decoupling system, and **f** the field profile in ordinary Anderson off-diagonal disorder system. The field amplitude in (**f**) is magnified (×100) for the presentation. Figure and caption adapted from Yu et al. (2017b) under a CC BY license (http://creativecommons.org/licenses/by/4.0/)

incoherent blockade of wave transport from Anderson localization (35 dB enhancement from 0.03% transmission at Fig. 3.14f). The eigenmode decoupling thus allows for the decoupling inside randomly distributed resonator systems, compensating the Anderson blockade from the off-diagonal disorder, as an example of the designer disorder (Yu et al. 2016a, d; Florescu et al. 2009; Torquato et al. 2015; Chertkov et al. 2016). Despite the requirement of the designed on-site energy distribution, the decoupling without any symmetry provides a novel route to "hiding" elements in coupled systems.

Is Target Decoupling Robust? We now estimate the fabrication errors in the distance between resonators and the radius of each resonator (Fig. 3.15a), to examine the stability of the eigenmode decoupling system. The distance is set to be $d_{err} = d \cdot (1 + \Delta_d \cdot u[-0.5, 0.5])$ and the radius of each resonator is set to be $r_{0\text{-err}} = r_0 \cdot (1 + \Delta_r \cdot u[-0.5, 0.5])$ where d and r_0 are the original distance and radius values, and Δ_d and Δ_r are the magnitude of fabrication errors, respectively. The result for each error is shown in Fig. 3.15b, c, presenting the transmission with respect to the magnitude of the fabrication errors Δ_d and Δ_r, and also comparing with the results from ordinary crystal systems. As seen, although the eigenmode decoupling system is more resistant to the distance error (or coupling error until ~5%), the stable

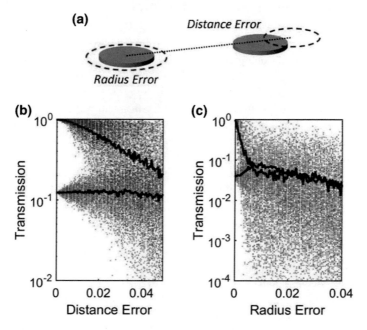

Fig. 3.15 Effect of fabrication errors. **a** A schematic of the fabrication errors on the resonator radius and the distance between resonators. **b, c** Transmissions of the decoupling system (blue) and ordinary crystal system (orange) for the **b** distance error and **c** resonator radius error. Each dot denotes the transmission for each case of fabrication errors. Solid lines denote the ensemble average of 100 random samples. Target decoupling systems are generally robust to both types of errors. Figure and caption adapted from Yu et al. (2017b) under a CC BY license (http://creativecommons.org/licenses/by/4.0/)

regime of the eigenmode decoupling system is fragile to the radius error (or on-site energy error, <0.7%). This phenomenon, the sensitivity to on-site energy errors, is an inherent property of weakly-coupled systems (Haus 1984), because the self-energy mismatch prohibits the complete energy transfer between optical elements.

Impact and Limit From the deterministic operation based on the eigenmode, the applications exploiting multimodal (Yu et al. 2013; Longhi 2016) or continuous (Yu et al. 2015b; Horsley et al. 2015; Longhi 2014) non-Hermitian potentials can also be envisaged for the defect-resistant realization of lasers or absorbers. Also note that the target decoupling technique, separating target elements from the other region in "wave" networks, also possesses the interdisciplinary link with the selective target control in network theory (Iudice et al. 2015; Gao et al. 2014; Skardal and Arenas 2015). However, likewise the global scattering increase in spectral domain in most of cloaking structures (Monticone and Alù 2013) (except few extreme cases: diamagnetic and superconducting cloaks (Monticone and Alù 2013)), the bandwidth problem in the target decoupling is the engineering subject which can be improved by alleviating the strict decoupling condition.

3.5 Generalization: Bohmian Photonics

In Sects. 3.1–3.4, we developed the concept of metadisorder. Based on the target control of an eigenmode using the degree of freedom in the on-site energy, the introduction of metadisorder structures enables error-robust optical functions with coherence (Figs. 3.7, 3.8, 3.9, 3.13, and 3.14 with Figs. 3.10, and 3.15), and the independent control of wave quantities: modal localization and wave transport (Fig. 3.6, small-world-like behaviour). In this section, especially generalizing the independent control method, we introduce the independent control of more fundamental wave quantities of light: amplitude and phase.

The information of a scalar wave, which has the form of $A(\mathbf{x})exp[i(\mathbf{k} \cdot \mathbf{x} + \varphi)]$, is defined by the functions of amplitude $A(\mathbf{x})$ and phase $\mathbf{k} \cdot \mathbf{x} + \varphi$. Could we access the complete information of each quantity independently? Actually, it is very difficult. While the scattering, or more broadly, the wave-matter interaction, is used to manipulate the amplitude information of the wave, the scattering simultaneously distorts the phase information according to the mathematical form $A(\mathbf{x})exp[i(\mathbf{k} \cdot \mathbf{x} + \varphi)]$ that shows the entangled amplitude and phase functions (Fig. 3.16, centre). This limit was also confirmed with the amplitude and phase uncertainty relation in quantum-mechanical systems (Louisell et al. 1963): closed systems with Hermitian Hamiltonians.

However, we can get insights from non-Hermitian physics (Bender and Boettcher 1998; Bender et al. 1999, 2002; Bender 2007) for open systems. For the limit of the least time for the state transition originating from the time-energy uncertainty principle, C. M. Bender et al. demonstrated that arbitrarily short time transitions can be achieved with the extension to non-Hermitian quantum mechanics (Bender et al.

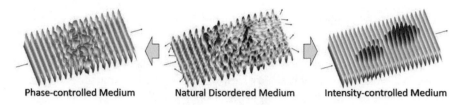

Phase-controlled Medium   Natural Disordered Medium   Intensity-controlled Medium

Fig. 3.16 Independent control of amplitude and phase information of light. (Centre) Simultaneous distortions of amplitude and phase distributions in disordered media. The examples of independently controlled amplitude and phase functions are shown: (Left) Constant-intensity light with freely-tunable phase information, and (Right) Vacuum-like phase evolution with strong light energy confinement

2007). From this viewpoint, also considering the emerging field of non-Hermitian photonics (Guo et al. 2009; Longhi 2009a; Rüter et al. 2010; Regensburger et al. 2012; Yu et al. 2012, 2013, 2015a, b, 2016b, c, 2017a, 2018b, c; Feng et al. 2013; Ge and Stone 2014; Hodaei et al. 2014; Peng et al. 2014; Assawaworrarit et al. 2017; Teimourpour et al. 2016, 2017a, b; Ge and El-Ganainy 2016; Lawrence et al. 2014; Piao et al. 2018; Yu et al. 2019a, b), we can envisage the independent control of amplitude and phase information (Fig. 3.16, left and right) using complex optical potentials.

For this purpose, we find a clue in quantum mechanics: the de Broglie-Bohm interpretation of quantum mechanics. We develop the design of complex potentials for the independent control of amplitude and phase information of light, by applying the Bohmian formulation of the electromagnetic wave equation.

Bohmian Mechanics Although quantum mechanics is founded upon rigorous experimental evidences, these evidences can be interpreted in different ways. The de Broglie-Bohm theory (Bohm 1952), also called Bohmian mechanics, is one of the alternative views on quantum mechanics, suggested by de Broglie in 1927 and rediscovered by Bohm (1952). Compared to the mainstream view, i.e., the Copenhagen interpretation that emphasizes the indeterministic nature of quantum phenomena (Heisenberg 1958), the focus of Bohmian mechanics is to understand "what is really happening at the quantum level" by tracking an individual event with the reintroduction of a "definite" particle position. This point-particle view allows the classical-like expectation of quantum particle trajectories (Bohm 1952; Holland et al. 1995), despite the ongoing debate (Englert et al. 1992, 1993; Dürr and Goldstein 1261; Scully 1998; Golshani and Akhavan 2001; Mahler et al. 2016): the "surreal" trajectory in a Welcher-Weg measurement for the individual measurement of entangled particles. Until now, Bohmian mechanics does not contradict the Copenhagen interpretation and quantum-mechanical experiments; recent achievements are discussed in Nassar (2017).

The heart of the Bohmian formulation is in the polar form wavefunction $\psi = Re^{i(S/\hbar)}$, composed of the "real-valued" functions of amplitude R and phase S. By applying this form to the Schrödinger equation $i\hbar\partial_t\psi = -(\hbar^2/2\,m)\nabla^2\psi + V\psi$ and then separating its real and imaginary parts (Holland et al. 1995), we achieve the (i) quantum Hamilton-Jacobi (QHJ) equation $\partial_t S + (\nabla S)^2/2\,m - (\hbar^2/2\,m)\nabla^2 R/R + V =$

0, and (ii) continuity equation $\partial_t R^2 + \nabla \cdot (R^2 \nabla S/m) = 0$, which identify an individual particle trajectory analogous to Newton's second law. The exerted force is divided by two parts: the classical force $-\nabla V$ and quantum force $(\hbar^2/2\,m)\nabla(\nabla^2 R/R)$. Same as the Copenhagen interpretation, the annihilation of the quantum force from $\hbar \to 0$ approximates quantum phenomena by Newtonian mechanics. Another point is that at least in the QHJ equation, the amplitude R and phase S are separated, while they are re-entangled in the continuity equation. Could we untie this knot? Non-Hermitian photonics presents the solution.

Bohmian Photonics We introduce the concept of Bohmian mechanics to non-Hermitian photonics. For the transverse electric (TE) mode, the field evolution in the x-y plane obeys the Helmholtz equation $\nabla^2 E_z + V(x, y)E_z = 0$, where $V = k_0^2 \varepsilon(x, y)$ is defined by a complex permittivity $\varepsilon(x, y) = \varepsilon_r(x, y) + i\varepsilon_i(x, y)$ and $k_0 = \omega/c$. We express the field profile in polar form (Holland 1995) as $E_z(x, y) = R(x, y)e^{iS(x,y)}$, where R is a real amplitude function that is by definition nonnegative, and S is a real phase function. The instantaneous wavevector is determined by $\mathbf{k}(x, y) = \nabla S$, which corresponds to the guidance equation (Holland 1995). Applying this polar-form field to the equation, and then separating its real and imaginary parts derive the governing equations for "Bohmian photonics"; for the complex optical potential $V = V_r(x, y) + iV_i(x, y)$, we achieve:

$$V_r(x, y) = |\nabla S|^2 - \frac{\nabla^2 R}{R}, \tag{3.13}$$

$$V_i(x, y) = -\nabla^2 S - 2\frac{\nabla R}{R} \cdot \nabla S. \tag{3.14}$$

Equation (3.13) corresponds to the photonic counterpart of the stationary ($\partial_t \to 0$) QHJ equation (Holland 1995). Similar to the QHJ equation in Bohmian mechanics, the real part of the optical potential $V_r = k_0^2 \varepsilon_r$ (Fig. 3.17a) is decomposed into the "classical" Hamilton-Jacobi term $V_{\text{classical}} = |\nabla S|^2$ (Fig. 3.17b) and the "quantum potential" counterpart (Fig. 3.17c) $V_{\text{quantum}} = -\nabla^2 R/R$. Each potential *separately* governs the spatial phase evolution with $S(x, y)$ (Fig. 3.17d) and the optical confinement with $R(x, y)$ (Fig. 3.17e), constructing the full spatial information of the wave as $E_z = Re^{iS}$ (Fig. 3.17f). Equation (3.14), rewritten as $\nabla \cdot (R^2 \nabla S) + k_0^2 \varepsilon_i(x, y)R^2 = 0$, is the stationary continuity equation, which describes the equilibrium between optical sources (or sinks) from $\varepsilon_i(x, y)$ and the divergence of the power flow $R^2 \nabla S$.

Although the Bohmian formulation is mathematically identical to the original Helmholtz wave equation, the Bohmian interpretation of the complex optical potential clarifies the route to the independent control of the phase and amplitude information of light, analogous to the separation of classical and quantum phenomena in Bohmian mechanics (Holland 1995). For the given field $E_z = Re^{iS}$ with targeted real functions of $S(x, y)$ and $R(x, y)$ that satisfy the electromagnetic boundary conditions, the real optical potential V_r is directly derived from the sum of $V_{\text{classical}} = |\nabla S|^2$ and $V_{\text{quantum}} = -\nabla^2 R/R$. The main difference between Bohmian photonics and Bohmian mechanics is in the presence of the imaginary optical potential V_i in non-Hermitian

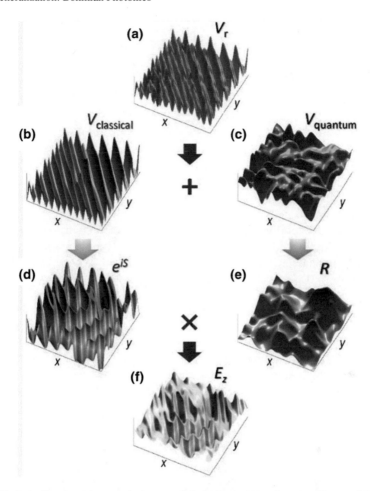

Fig. 3.17 Bohmian photonics: **a** V_r is the sum of the **b** classical Hamilton-Jacobi potential $V_{classical}$ $= |\nabla S|^2$ and **c** quantum potential $V_{quantum} = -\nabla^2 R/R$, each governing the **d** phase evolution e^{iS} and **e** amplitude R of the **f** entire field E_z. Figure and caption adapted from Yu et al. (2018b) under a CC BY license (http://creativecommons.org/licenses/by/4.0/)

Photonics, which satisfies the necessary degrees of freedom for the simultaneous control of amplitude and phase functions. V_i is determined by the continuity equation Eq. (3.14), eventually accomplishing the inverse design of the complex potential V for the given field $E_z = Re^{iS}$.

Bohmian photonics also offers an interesting perspective on "how to understand zero refractive index". In the field of metamaterials, near-zero index materials have been interpreted as the material extremely "stretching" the wavelength of light (Liberal and Engheta 2017), from the definition of the effective wavelength $\lambda_{eff} = \lambda_0/n$. In Bohmian photonics, if we set $\nabla R = \nabla S = 0$, the constant amplitude and phase, we achieve the zero index from $V_r = V_i = 0$ and thus $\varepsilon_r = \varepsilon_i = 0$. This view with the alleviation of the strict condition $\nabla R = \nabla S = 0$ allows the extension of near-zero refractive index photonics (Liberal and Engheta 2017) to the non-Hermitian regime: amplitude-conserved or phase-conserved materials. We now introduce two examples for the independent control of amplitude and phase information of light in disordered photonics: (i) amplitude-conserved metadisorder and (ii) phase-conserved metadisorder.

Amplitude-Conserved Metadisorder First, we explore a particular case: quantum-mechanically-free (Holland 1995) potentials of $V_{\text{quantum}} = -\nabla^2 R/R = 0$. Among the general solutions of Laplace's equation $\nabla^2 R = 0$, we focus on the constant-intensity (CI) wave (Makris et al. 2015, 2017; Rivet et al. 2017, 2018) with $R(x, y) = R_0$. The optical potential for the CI wave then becomes $V_r(x, y) = |\nabla S|^2$ and $V_i(x, y) = -\nabla^2 S$, providing the complex permittivity profile of $\varepsilon(x, y) = (|\nabla S|^2 - i\nabla^2 S)/k_0^2$. This relation between the material profile and phase function enables the alternative classification of "homogeneous" materials to include the following: (i) ordinary material for the linearly varying phase $S(x, y) = k_x x + k_y y$ with $\varepsilon(x, y) = (k_x^2 + k_y^2)/k_0^2 > 0$ and (ii) epsilon-near-zero (ENZ) material for the constant phase $S(x, y) = S_0$ with $\varepsilon(x, y) = 0$. Because $Re[\varepsilon] = |\nabla S|^2/k_0^2 \geq 0$, this CI-wave class does not include negative real permittivity. Depending on the form of $S(x, y)$, or the detailed spatial distribution of light waves, the CI-wave class can also include "inhomogeneous" materials with non-constant first-order partial derivatives of $S(x, y)$.

Figure 3.18 shows an example of CI-wave propagation in a 2D disordered potential, called an "amplitude-conserved metadisorder". We design the perturbed phase function $S(x, y)$ from the plane wave-like linear phase evolution, with $S(x, y) = n_0 k_0 y$

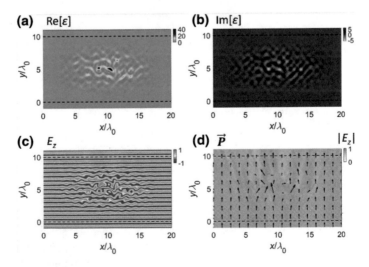

Fig. 3.18 Amplitude-conserved Bohmian metadisorder. **a** Real and **b** imaginary permittivity distributions and **c** the obtained field evolution. **d** The distribution of the Poynting vector overlaid on the amplitude of the electric field E_z. Figure and caption adapted from Yu et al. (2018b) under a CC BY license (http://creativecommons.org/licenses/by/4.0/)

$+ 2\pi W(x, y)\Delta S(x, y)/\max\{W(x, y)\Delta S(x, y)\}$ for the background refractive index n_0, where $\Delta S(x, y)$ is the random perturbation defined in the spectral domain as

$$\Delta S(x, y) = u[-1, 1] \int_0^{k_0} \sin(q_x x + u[-\pi, \pi])dq_x + u[-1, 1] \int_0^{k_0} \sin(q_y y + u[-\pi, \pi])dq_y, \quad (3.15)$$

$u[a, b]$ is a uniform random number between a and b. $W(x, y)$ is the weighting function for the electromagnetic continuity condition of $S(x, y)$ given by

$$W = \left[1 - \cos\left(\frac{2\pi x}{L_x}\right)\right]\left[1 - \cos\left(\frac{2\pi y}{L_y}\right)\right], \quad (3.16)$$

where L_x and L_y are the lengths of each side of the rectangle-shaped design area. From the designed $S(x, y)$, the necessary permittivity distribution is obtained as $\varepsilon(x, y) = (|\nabla S|^2 - i\nabla^2 S)/k_0^2$ (Fig. 3.18a, b). Figure 3.18c shows the result of independent phase control from the CI wave condition, a random phase in the design area and a perfect plane wave at the input and output regions without any scattering. The Poynting vector distribution (Fig. 3.18d) shows that this CI-wave propagation is obtained from the modified magnetic field while preserving the intensity of the electric field. Notably, this Bohmian random potential supporting the CI wave with tunable phase evolution belongs to the class of metadisorder as defined in Sect. 3.1.

Phase-Conserved Metadisorder We use the nonzero quantum potential $V_{\text{quantum}} = -\nabla^2 R/R \neq 0$, which allows for the realization of exotic energy confinement and annihilation. The confinement of light is achieved with photonic structures that hinder outgoing waves (Joannopoulos et al. 2011; Lagendijk et al. 2009; Lee et al. 2012), except for a few examples, such as the use of bound states in the continuum (Marinica et al. 2008; Hsu et al. 2013, 2016). Because all of these methods are based on the scattering of waves, the confinement of light, or, more generally, the modulation of the amplitude function R, leads to a subsequent disturbance to the phase function S. Instead, we apply the Bohmian approach to achieve "scattering-free" energy confinement or annihilation, perfectly preserving the phase information.

Without loss of generality, we assume a linear phase evolution $S(x, y) = n_0 k_0 y$ in the entire space. The required potential then becomes

$$V_r(x, y) = n_0^2 k_0^2 - \frac{\nabla^2 R}{R}$$
$$V_i(x, y) = -2n_0 k_0 \frac{\nabla R}{R} \cdot \hat{y} \quad (3.17)$$

To achieve the scattering-free condition, $\nabla R/R$ and $\nabla^2 R/R$ should be continuous, requiring the R function to be of the C^2 differentiability class. As an example, we employ a "bump" function that is smooth and has continuous derivatives of all the orders, i.e., of C^∞ class. If we assume the design of metadisorder structures that preserve the linear phase evolution $S(x, y) = n_0 k_0 y$, the spatial profile of the potential can be obtained from the superposition of the bump functions. The amplitude function

$R(x, y)$ is defined as $R(x, y) = 1 + \sum_{j=1}^{N} B_j(x, y)$, where the form of the jth bump function $B_j(x, y)$ is

$$B_j(x, y) = \begin{cases} R_0 \exp\left(-\frac{b^2\lambda_0^2}{\sigma_j^2-r_j^2}\right) & \text{for } r_j \leq \sigma_j \\ 0 & \text{otherwise} \end{cases}, \tag{3.18}$$

where R_0 determines the strength of energy confinement, b represents the sharpness of the confinement function, $\sigma_j = 2\lambda_0 + \lambda_0 u[-1, 1]$ defines the confinement area in the polar coordinate representation of $r_j = [(x-x_{j0})^2 + (y-y_{j0})^2]^{1/2}$ for (x_{j0}, y_{j0}) $= (10\lambda_0 + \lambda_0 u[-1, 1], 10\lambda_0 + \lambda_0 u[-1, 1])$ with $u[a_1, a_2]$, which is a uniform random number between a_1 and a_2. This mathematical form of $R(x, y)$ provides the spatial profile of the amplitude, which is differentiable and disordered, leading to the phase-conserved metadisorder structure.

Anomalous confinement or annihilation of optical energy that perfectly preserves the phase information of light is then demonstrated with the construction of a scattering-free quantum-like potential defined by $R(x, y)$. Figure 3.19 shows the examples of energy confinement (Fig. 3.19c, d) and cancellation (Fig. 3.19g, h) of light waves with random profiles. The wave propagations are calculated with the permittivity profiles of the superposition of the bump functions $R(x, y)$ (Fig. 3.19a, b, e, f). The incident phase information is perfectly preserved in all cases, despite highly disordered potential profiles and different amplitude distributions.

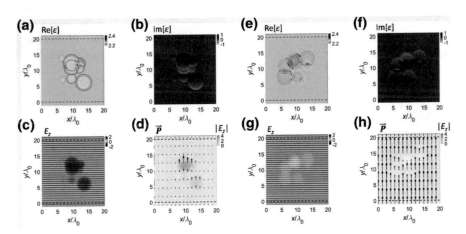

Fig. 3.19 Phase-conserved Bohmian metadisorder. **a–d** Energy confinement and **e–h** annihilation in the designed regions. **a, e** Real and **b, f** imaginary permittivity distributions and **c, g** obtained field evolutions. **d, h** The distributions of the Poynting vector overlaid on the amplitude of the electric field E_z. The confinement and cancellation of optical energy are obtained with **a–d** $R_0 = 2$ and **e–h** $R_0 = -0.6$ for Eq. (3.18), respectively. $n_0 = 1.5$, and $b = 2$ for defining the 2D bump functions. Figure and caption adapted from Yu et al. (2018b) under a CC BY license (http://creativecommons. org/licenses/by/4.0/)

Compared with conventional transformation optics (Pendry et al. 2006) and the spatial Kramers-Kronig relation (Horsley et al. 2015), Bohmian photonics provides a new design strategy for artificial inhomogeneous media considering the independent handling of wave quantities. The inverse design technique for the given phase and amplitude distributions can be applied to metamaterial design, for largely positive (Choi et al. 2011), zero (Liberal and Engheta 2017), and negative (Smith et al. 2004) real values and imaginary values (Rüter et al. 2010; Hodaei et al. 2014; Peng et al. 2014; Assawaworrarit et al. 2017; Chang et al. 2014; Feng et al. 2014) of material parameters. The independent control of S and R enables the intuitive design of antennas and resonators, respectively, especially in a non-Hermitian regime. The application of the alternative interpretation in quantum mechanics also provides a new perspective in the field of quantum-optical analogy (Longhi 2009b).

References

Amaral, L.A.N., Scala, A., Barthelemy, M., Stanley, H.E.: Classes of small-world networks. Proc. Natl. Acad. Sci. **97**, 11149–11152 (2000)

Anderson, P.W.: Absence of diffusion in certain random lattices. Phys. Rev. **109**, 1492 (1958)

Aspuru-Guzik, A., Walther, P.: Photonic quantum simulators. Nat. Phys. **8**, 285–291 (2012)

Assawaworrarit, S., Yu, X., Fan, S.: Robust wireless power transfer using a nonlinear parity-time-symmetric circuit. Nature **546**, 387–390 (2017)

Bender, C.M.: Making sense of non-Hermitian Hamiltonians. Rep. Prog. Phys. **70**, 947–1018 (2007). https://doi.org/10.1088/0034-4885/70/6/r03

Bender, C.M., Boettcher, S.: Real spectra in non-Hermitian Hamiltonians having PT symmetry. Phys. Rev. Lett. **80**, 5243 (1998)

Bender, C.M., Boettcher, S., Meisinger, P.N.: PT-symmetric quantum mechanics. J. Math. Phys. **40**, 2201–2229 (1999)

Bender, C.M., Brody, D.C., Jones, H.F.: Complex extension of quantum mechanics. Phys. Rev. Lett. **89**, 270401 (2002)

Bender, C.M., Brody, D.C., Jones, H.F., Mcister, B.K.: Faster than Hermitian quantum mechanics. Phys. Rev. Lett. **98**, 040403 (2007)

Bohm, D.: A suggested interpretation of the quantum theory in terms of "hidden" variables. I. Phys. Rev. **85**, 166 (1952)

Bullmore, E.T., Bassett, D.S.: Brain graphs: graphical models of the human brain connectome. Annu. Rev. Clin. Psychol. **7**, 113–140 (2011)

Cai, W., Chettiar, U.K., Kildishev, A.V., Shalaev, V.M.: Optical cloaking with metamaterials. Nat. Photonics **1**, 224–227 (2007)

Chang, L., Jiang, X., Hua, S., Yang, C., Wen, J., Jiang, L., Li, G., Wang, G., Xiao, M.: Parity-time symmetry and variable optical isolation in active–passive-coupled microresonators. Nat. Photonics **8**, 524–529 (2014)

Chen, H., Chan, C., Sheng, P.: Transformation optics and metamaterials. Nat. Mater. **9**, 387–396 (2010)

Chertkov, E., DiStasio Jr., R.A., Zhang, G., Car, R., Torquato, S.: Inverse design of disordered stealthy hyperuniform spin chains. Phys. Rev. B **93**, 064201 (2016)

Choi, M., Lee, S.H., Kim, Y., Kang, S.B., Shin, J., Kwak, M.H., Kang, K.-Y., Lee, Y.-H., Park, N., Min, B.: A terahertz metamaterial with unnaturally high refractive index. Nature **470**, 369–373 (2011)

Christodoulides, D.N., Lederer, F., Silberberg, Y.: Discretizing light behaviour in linear and non-linear waveguide lattices. Nature **424**, 817–823 (2003)

Dürr, D., Fusseder, Goldstein, S., Zanghi, N.: Comment on "Surrealistic Bohm trajectories". Z. Naturforsch. A **48**, 1261 (1993). https://doi.org/10.1515/zna-1993-1219

El-Ganainy, R., Makris, K.G., Khajavikhan, M., Musslimani, Z.H., Rotter, S., Christodoulides, D.N.: Non-Hermitian physics and PT symmetry. Nat. Phys. **14**, 11–19 (2018)

Englert, B.-G., Scully, M.O., Süssmann, G., Walther, H.: Surrealistic Bohm trajectories. Z. Naturforsch. A **47**, 1175–1186 (1992)

Englert, B.-G., Scully, M., Süssmann, G., Walther, H.: Reply to comment on "Surrealistic Bohm trajectories". Z. Naturforsch. A **48**, 1263–1264 (1993)

Feng, L., Xu, Y.L., Fegadolli, W.S., Lu, M.H., Oliveira, J.E., Almeida, V.R., Chen, Y.F., Scherer, A.: Experimental demonstration of a unidirectional reflectionless parity-time metamaterial at optical frequencies. Nat. Mater. **12**, 108–113 (2013). https://doi.org/10.1038/nmat3495

Feng, L., Wong, Z.J., Ma, R.-M., Wang, Y., Zhang, X.: Single-mode laser by parity-time symmetry breaking. Science **346**, 972–975 (2014)

Feng, L., El-Ganainy, R., Ge, L.: Non-Hermitian photonics based on parity-time symmetry. Nat. Photonics **11**, 752–762 (2017)

Florescu, M., Torquato, S., Steinhardt, P.J.: Designer disordered materials with large, complete photonic band gaps. Proc. Natl. Acad. Sci. U.S.A. **106**, 20658–20663 (2009)

Gao, J., Liu, Y.-Y., D'Souza, R.M., Barabási, A.-L.: Target control of complex networks. Nat. Commun. **5**, 5415 (2014). https://doi.org/10.1038/ncomms6415

Garanovich, I.L., Longhi, S., Sukhorukov, A.A., Kivshar, Y.S.: Light propagation and localization in modulated photonic lattices and waveguides. Phys. Rep. **518**, 1–79 (2012)

Ge, L., El-Ganainy, R.: Nonlinear modal interactions in parity-time (PT) symmetric lasers. Sci. Rep. **6**, 24889 (2016)

Ge, L., Stone, A.D.: Parity-time symmetry breaking beyond one dimension: the role of degeneracy. Phys. Rev. X **4**, 031011 (2014). https://doi.org/10.1103/PhysRevX.4.031011

Golshani, M., Akhavan, O.: Bohmian prediction about a two double-slit experiment and its disagreement with standard quantum mechanics. J. Phys. A **34**, 5259 (2001)

Guo, A., Salamo, G.J., Duchesne, D., Morandotti, R., Volatier-Ravat, M., Aimez, V., Siviloglou, G.A., Christodoulides, D.N.: Observation of PT-symmetry breaking in complex optical potentials. Phys. Rev. Lett. **103** (2009). https://doi.org/10.1103/physrevlett.103.093902

Hahn, K., Kärger, J., Kukla, V.: Single-file diffusion observation. Phys. Rev. Lett. **76**, 2762 (1996)

Haus, H.A.: Waves and fields in optoelectronics. In: Solid State Physical Electronics. Prentice-Hall, Englewood Cliffs, NJ (1984)

Heisenberg, W.: Physics and Philosophy: The Revolution in Modern Science. George Allen & Unwin Ltd., London (1958)

Hilgetag, C.C., Goulas, A.: Is the brain really a small-world network? Brain Struct. Funct. **221**, 2361–2366 (2016)

Hodaei, H., Miri, M.-A., Heinrich, M., Christodoulides, D.N., Khajavikhan, M.: Parity-time-symmetric microring lasers. Science **346**, 975–978 (2014)

Holland, P.R.: The Quantum Theory of Motion: An Account of the De Broglie-Bohm Causal Interpretation of Quantum Mechanics. Cambridge University Press (1995)

Horsley, S., Artoni, M., La Rocca, G.: Spatial Kramers-Kronig relations and the reflection of waves. Nat. Photonics **9**, 436–439 (2015)

Hsieh, P., Chung, C., McMillan, J., Tsai, M., Lu, M., Panoiu, N., Wong, C.W.: Photon transport enhanced by transverse Anderson localization in disordered superlattices. Nat. Phys. **11**, 268 (2015)

Hsu, C.W., Zhen, B., Lee, J., Chua, S.L., Johnson, S.G., Joannopoulos, J.D., Soljacic, M.: Observation of trapped light within the radiation continuum. Nature **499**, 188–191 (2013). https://doi.org/10.1038/nature12289

Hsu, C.W., Zhen, B., Stone, A.D., Joannopoulos, J.D., Soljačić, M.: Bound states in the continuum. Nat. Rev. Mater. **1**, 16048 (2016)

Huang, X., Lai, Y., Hang, Z.H., Zheng, H., Chan, C.: Dirac cones induced by accidental degeneracy in photonic crystals and zero-refractive-index materials. Nat. Mater. **10**, 582–586 (2011)

Iudice, F.L., Garofalo, F., Sorrentino, F.: Structural permeability of complex networks to control signals. Nat. Commun. **6**, 8349 (2015). https://doi.org/10.1038/ncomms9349

Joannopoulos, J.D., Johnson, S.G., Winn, J.N., Meade, R.D.: Photonic Crystals: Molding the Flow of Light. Princeton University Press (2011)

Keil, R., Pressl, B., Heilmann, R., Gräfe, M., Weihs, G., Szameit, A.: Direct measurement of second-order coupling in a waveguide lattice (2015). arXiv:1510.07900

Lagendijk, A., Van Tiggelen, B., Wiersma, D.S.: Fifty years of Anderson localization. Phys. Today **62**, 24–29 (2009)

Lahini, Y., Avidan, A., Pozzi, F., Sorel, M., Morandotti, R., Christodoulides, D.N., Silberberg, Y.: Anderson localization and nonlinearity in one-dimensional disordered photonic lattices. Phys. Rev. Lett. **100**, 013906 (2008)

Lawrence, M., Xu, N., Zhang, X., Cong, L., Han, J., Zhang, W., Zhang, S.: Manifestation of PT symmetry breaking in polarization space with terahertz metasurfaces. Phys. Rev. Lett. **113**, 093901 (2014)

Lee, H., Chen, T., Li, J., Yang, K.Y., Jeon, S., Painter, O., Vahala, K.J.: Chemically etched ultrahigh-Q wedge-resonator on a silicon chip. Nat. Photonics **6**, 369–373 (2012)

Liang, G., Chong, Y.: Optical resonator analog of a two-dimensional topological insulator. Phys. Rev. Lett. **110**, 203904 (2013)

Liberal, I., Engheta, N.: Near-zero refractive index photonics. Nat. Photonics **11**, 149–158 (2017). https://doi.org/10.1038/nphoton.2017.13

Lidorikis, E., Sigalas, M., Economou, E.N., Soukoulis, C.: Tight-binding parametrization for photonic band gap materials. Phys. Rev. Lett. **81**, 1405 (1998)

Lin, Z., Ramezani, H., Eichelkraut, T., Kottos, T., Cao, H., Christodoulides, D.N.: Unidirectional invisibility induced by PT-symmetric periodic structures. Phys. Rev. Lett. **106**, 213901 (2011). https://doi.org/10.1103/PhysRevLett.106.213901

Longhi, S.: Bloch oscillations in complex crystals with PT symmetry. Phys. Rev. Lett. **103** (2009a). https://doi.org/10.1103/physrevlett.103.123601

Longhi, S.: Quantum-optical analogies using photonic structures. Laser Photon. Rev. **3** 243–261 (2009b)

Longhi, S.: Invisibility in PT-symmetric complex crystals. J. Phys. A **44**, 485302 (2011). https://doi.org/10.1088/1751-8113/44/48/485302

Longhi, S.: Talbot self-imaging in PT-symmetric complex crystals. Phys. Rev. A **90**, 043827 (2014)

Longhi, S.: PT phase control in circular multi-core fibers. Opt. Lett. **41**, 1897–1900 (2016)

Longhi, S., Della Valle, G., Ornigotti, M., Laporta, P.: Coherent tunneling by adiabatic passage in an optical waveguide system. Phys. Rev. B **76**, 201101 (2007)

Louisell, W.: Amplitude and phase uncertainty relations. Phys. Lett. **7** (1963)

Mahler, D.H., Rozema, L., Fisher, K., Vermeyden, L., Resch, K.J., Wiseman, H.M., Steinberg, A.: Experimental nonlocal and surreal Bohmian trajectories. Sci. Adv. **2**, e1501466 (2016)

Makris, K.G., Musslimani, Z.H., Christodoulides, D.N., Rotter, S.: Constant-intensity waves and their modulation instability in non-Hermitian potentials. Nat. Commun. **6**, 7257 (2015)

Makris, K.G., Brandstötter, A., Ambichl, P., Musslimani, Z.H., Rotter, S.: Wave propagation through disordered media without backscattering and intensity variations. Light Sci. Appl. **6**, e17035 (2017)

Marinica, D., Borisov, A., Shabanov, S.: Bound states in the continuum in photonics. Phys. Rev. Lett. **100**, 183902 (2008)

Martin, L., Di Giuseppe, G., Perez-Leija, A., Keil, R., Dreisow, F., Heinrich, M., Nolte, S., Szameit, A., Abouraddy, A.F., Christodoulides, D.N.: Anderson localization in optical waveguide arrays with off-diagonal coupling disorder. Opt. Express **19**, 13636–13646 (2011)

Molina-Terriza, G., Torres, J.P., Torner, L.: Twisted photons. Nat. Phys. **3**, 305–310 (2007)

Monticone, F., Alù, A.: Do cloaked objects really scatter less? Phys. Rev. X **3**, 041005 (2013)

Mrejen, M., Suchowski, H., Hatakeyama, T., Wu, C., Feng, L., O'Brien, K., Wang, Y., Zhang, X.: Adiabatic elimination-based coupling control in densely packed subwavelength waveguides. Nat. Commun. **6** (2015a)

Mrejen, M., Suchowski, H., Hatakeyama, T., Wu, C., Feng, L., O'Brien, K., Wang, Y., Zhang, X.: Adiabatic elimination-based coupling control in densely packed subwavelength waveguides. Nat. Commun. **6**, 7565 (2015b). https://doi.org/10.1038/ncomms8565

Nagarajan, R., Joyner, C.H., Schneider Jr., R.P., Bostak, J.S., Butrie, T., Dentai, A.G., Dominic, V.G., Evans, P.W., Kato, M., Kauffman, M.: Large-scale photonic integrated circuits. IEEE J. Sel. Top. Quantum Electron. **11**, 50–65 (2005)

Nassar, A.B., Miret-Artés, S.: Bohmian Mechanics, Open Quantum Systems and Continuous Measurements. Springer (2017)

Papasimakis, N., Fedotov, V.A., Fu, Y.H., Tsai, D.P., Zheludev, N.I.: Coherent and incoherent metamaterials and order-disorder transitions. Phys. Rev. B **80**, 041102 (2009)

Pendry, J.: Off-diagonal disorder and 1D localisation. J. Phys. C **15**, 5773 (1982)

Pendry, J.B., Schurig, D., Smith, D.R.: Controlling electromagnetic fields. Science **312**, 1780–1782 (2006)

Peng, B., Özdemir, Ş.K., Lei, F., Monifi, F., Gianfreda, M., Long, G.L., Fan, S., Nori, F., Bender, C.M., Yang, L.: parity-time-symmetric whispering-gallery microcavities. Nat. Phys. **10**, 394–398 (2014)

Piao, X., Yu, S., Park, N.: Design of transverse spinning of light with globally unique handedness. Phys. Rev. Lett. **120**, 203901 (2018)

Poddubny, A.N., Rybin, M.V., Limonov, M.F., Kivshar, Y.S.: Fano interference governs wave transport in disordered systems. Nat. Commun. **3**, 914 (2012)

Regensburger, A., Bersch, C., Miri, M.A., Onishchukov, G., Christodoulides, D.N., Peschel, U.: Parity-time synthetic photonic lattices. Nature **488**, 167–171 (2012). https://doi.org/10.1038/nature11298

Rivet, E., Brandstötter, A., Makris, K., Rotter, S., Lissek, H., Fleury, R.: Constant amplitude sound waves in non-Hermitian metamaterials. J. Acoust. Soc. Am. **142**, 2684–2684 (2017)

Rivet, E., Brandstötter, A., Makris, K.G., Lissek, H., Rotter, S., Fleury, R.: Constant-pressure sound waves in non-Hermitian disordered media. Nat. Phys. **14**, 942 (2018)

Rüter, C.E., Makris, K.G., El-Ganainy, R., Christodoulides, D.N., Segev, M., Kip, D.: Observation of parity-time symmetry in optics. Nat. Phys. **6**, 192–195 (2010)

Schwartz, T., Bartal, G., Fishman, S., Segev, M.: Transport and Anderson localization in disordered two-dimensional photonic lattices. Nature **446**, 52–55 (2007)

Scully, M.O.: Do Bohm trajectories always provide a trustworthy physical picture of particle motion? Phys. Scripta **76**, 41–46 (1998)

Shafiei, F., Monticone, F., Le, K.Q., Liu, X.-X., Hartsfield, T., Alù, A., Li, X.: A subwavelength plasmonic metamolecule exhibiting magnetic-based optical Fano resonance. Nat. Nanotech. **8**, 95–99 (2013)

Skardal, P.S., Arenas, A.: Control of coupled oscillator networks with application to microgrid technologies. Sci. Adv. **1**, e1500339 (2015)

Smith, D.R., Pendry, J.B., Wiltshire, M.C.: Metamaterials and negative refractive index. Science **305**, 788–792 (2004)

Song, W., Gatdula, R., Abbaslou, S., Lu, M., Stein, A., Lai, W.Y., Provine, J., Pease, R.F.W., Christodoulides, D.N., Jiang, W.: High-density waveguide superlattices with low crosstalk. Nat. Commun. **6**, 7027 (2015). https://doi.org/10.1038/ncomms8027

Takesue, H., Matsuda, N., Kuramochi, E., Munro, W.J., Notomi, M.: An on-chip coupled resonator optical waveguide single-photon buffer. Nat. Commun. **4**, 2725 (2013). https://doi.org/10.1038/ncomms3725

Teimourpour, M.H., Ge, L., Christodoulides, D.N., El-Ganainy, R.: Non-Hermitian engineering of single mode two dimensional laser arrays. Sci. Rep. **6**, 33253 (2016)

Teimourpour, M., Khajavikhan, M., Christodoulides, D.N., El-Ganainy, R.: Robustness and mode selectivity in parity-time (PT) symmetric lasers. Sci. Rep. **7**, 10756 (2017a)

Teimourpour, M., Rahman, A., Srinivasan, K., El-Ganainy, R.: Non-Hermitian engineering of synthetic saturable absorbers for applications in photonics. Phys. Rev. Appl. **7**, 014015 (2017b)

Torquato, S., Zhang, G., Stillinger, F.: Ensemble theory for stealthy hyperuniform disordered ground states. Phys. Rev. X **5**, 021020 (2015). https://doi.org/10.1103/PhysRevX.5.021020

Watts, D.J., Strogatz, S.H.: Collective dynamics of 'small-world' networks. Nature **393**, 440–442 (1998a)

Watts, D.J., Strogatz, S.H.: Collective dynamics of 'small-world' networks. Nature **393**, 440–442 (1998b)

Wiersma, D.S.: Disordered photonics. Nat. Photonics **7**, 188–196 (2013)

Wiersma, D.S., Bartolini, P., Lagendijk, A., Righini, R.: Localization of light in a disordered medium. Nature **390**, 671–673 (1997)

Xu, L., Chen, H.: Conformal transformation optics. Nat. Photonics **9**, 15–23 (2015)

Yu, S., Piao, X., Mason, D.R., In, S., Park, N.: Spatiospectral separation of exceptional points in PT-symmetric optical potentials. Phys. Rev. A **86**, 031802 (2012)

Yu, S., Mason, D.R., Piao, X., Park, N.: Phase-dependent reversible nonreciprocity in complex metamolecules. Phys. Rev. B **87**, 125143 (2013)

Yu, S., Piao, X., Hong, J., Park, N.: Progress toward high-Q perfect absorption: a Fano antilaser. Phys. Rev. A **92**, 011802 (2015a)

Yu, S., Piao, X., Yoo, K., Shin, J., Park, N.: One-way optical modal transition based on causality in momentum space. Opt. Express **23**, 24997–25008 (2015b)

Yu, S., Piao, X., Hong, J., Park, N.: Bloch-like waves in random-walk potentials based on supersymmetry. Nat. Commun. **6**, 8269 (2015c)

Yu, S., Piao, X., Hong, J., Park, N.: Metadisorder for designer light in random systems. Sci. Adv. **2**, e1501851 (2016a)

Yu, S., Park, H.S., Piao, X., Min, B., Park, N.: Low-dimensional optical chirality in complex potentials. Optica **3**, 1025 (2016b). https://doi.org/10.1364/OPTICA.3.001025

Yu, S., Piao, X., Park, N.: Acceleration toward polarization singularity inspired by relativistic E × B drift. Sci. Rep. **6**, 37754 (2016c)

Yu, S., Piao, X., Hong, J., Park, N.: Interdimensional optical isospectrality inspired by graph networks. Optica **3**, 836–839 (2016d)

Yu, S., Piao, X., Park, N.: Dynamical phase diagram of parity-time symmetry with competing saturable channels (2017a). arXiv:1707.07096

Yu, S., Piao, X., Park, N.: Target decoupling in coupled systems resistant to random perturbation. Sci. Rep. **7**, 2139 (2017b)

Yu, S., Piao, X., Park, N.: Disordered potential landscapes for anomalous delocalization and superdiffusion of light. ACS Photonics **5**, 1499 (2018a)

Yu, S., Piao, X., Park, N.: Bohmian photonics for independent control of the phase and amplitude of waves. Phys. Rev. Lett. **120**, 193902 (2018b)

Yu, S., Piao, X., Park, N.: Designing non-Hermitian dynamics for conservative state evolution on the Bloch sphere. Phys. Rev. A **97**, 033805 (2018c)

Yu, S., Piao, X., Park, N.: Chirality in non-Hermitian photonics (2019a). arXiv:1904.11662v2

Yu, S., Piao, X., Park, N.: Neuromorphic functions of light in parity-time-symmetric systems. Adv. Sci. **6**, 1900771 (2019b)

Zhu, X., Feng, L., Zhang, P., Yin, X., Zhang, X.: One-way invisible cloak using parity-time symmetric transformation optics. Opt. Lett. **38**, 2821–2824 (2013). https://doi.org/10.1364/OL.38.002821

Chapter 4
Conclusion and Outlook

In Chaps. 2 and 3, we investigated the independent controls of wave quantities in disordered photonic structures, revealing the existence of anomalous types of disorder that are distinct from both order and randomness. The proposed structures enable

I. Anomalous spectral responses: perfect bandgaps with modal localizations, random wave switching for binary and fuzzy logics, spectral correspondence between graphs and photonic structures, and interdimensional wave transport, and
II. Anomalous modal responses: metadisorder for small-world-like wave phenomena and robust signal transport, target decoupling for the cloaking inside coupled systems, and Bohmian photonics for the independent control of the amplitude and phase of light.

Achieving the advantages of order and randomness in the designed photonic disordered structures enables new types of wave phenomena, such as perfect bandgaps with energy confinement and coherent and error-robust guided waves, which construct toolkits for optical signal processing (Viciani et al. 2015; Shcherbakov et al. 2015; Yu et al. 2008, 2009, 2011, 2012, 2015; Yu and Fan 2009). The theory of Bohmian photonics (Yu et al. 2018a, b, c) also provides theoretical tools for the customization of light flows in non-Hermitian photonics (Feng et al. 2017; El-Ganainy et al. 2018).

The examples described in this book mainly focus on the control of wave quantities, including frequency ω, wavevector $\mathbf{k}$, and field profile $A(\mathbf{x})$. Therefore, we can envisage the extension of the proposed concept to other wave quantities and physical axes, including spin (Piao et al. 2018a; Bliokh et al. 2014, 2015; Bliokh and Nori 2012; Bekshaev et al. 2015; Bliokh et al. 2015; Yu et al. 2016a, b, 2018c; Piao et al. 2018a, b) and orbital (Bliokh et al. 2014; Vitullo et al. 2017; Allen et al. 1992; Mirhosseini et al. 2013; Karimi et al. 2014) angular momenta for 3D physical axes and the topological properties of light. In particular, quantization of the topology of band structures (Hasan and Kane 2010; Lu et al. 2014; Khanikaev et al. 2013; Rechtsman et al. 2013; Lustig et al. 2019) and k-space (Piao et al. 2018a; Krishnamoorthy et al. 2012) would provide a new perspective for understanding and manipulating the intermediate areas between order and randomness, such as the

© The Author(s), under exclusive license to Springer Nature Singapore Pte. Ltd. 2019
S. Yu et al., *Top-Down Design of Disordered Photonic Structures*,
SpringerBriefs in Physics, https://doi.org/10.1007/978-981-13-7527-9_4

realization of topological phenomena in disordered structures. In addition, time-axis expansion of disordered photonics would impose a new degree of freedom on time crystals (Zhang et al. 2017; Xu and Wu 2018; Lustig et al. 2018).

Our approaches are based on the use of concepts and methodologies from other fields, including supersymmetry, parity-time symmetry, Anderson localization in quantum mechanics, neuromorphic photonics, and small-world graphs in network theory, and their inspiration to the concept of metadisorder. In terms of methodologies, we envisage the application of machine learning techniques (Cubuk et al. 2015) to explore these complex, ambiguous, and nonlinear intermediate areas between order and randomness.

References

Allen, L., Beijersbergen, M.W., Spreeuw, R., Woerdman, J.: Orbital angular momentum of light and the transformation of Laguerre-Gaussian laser modes. Phys. Rev. A **45**, 8185 (1992)

Bekshaev, A.Y., Bliokh, K.Y., Nori, F.: Transverse spin and momentum in two-wave interference. Phys. Rev. X **5**, 011039 (2015)

Bliokh, K.Y., Nori, F.: Transverse spin of a surface polariton. Phys. Rev. A **85**, 061801 (2012)

Bliokh, K.Y., Bekshaev, A.Y., Nori, F.: Extraordinary momentum and spin in evanescent waves. Nat. Commun. **5**, 3300 (2014a). https://doi.org/10.1038/ncomms4300

Bliokh, K.Y., Dressel, J., Nori, F.: Conservation of the spin and orbital angular momenta in electromagnetism. New J. Phys. **16**, 093037 (2014b)

Bliokh, K.Y., Rodríguez-Fortuño, F., Nori, F., Zayats, A.V.: Spin-orbit interact of light. Nat. Photonics **9**, 796–808 (2015a)

Bliokh, K.Y., Smirnova, D., Nori, F.: Quantum spin Hall effect of light. Science **348**, 1448–1451 (2015b)

Cubuk, E.D., Schoenholz, S.S., Rieser, J.M., Malone, B.D., Rottler, J., Durian, D.J., Kaxiras, E., Liu, A.J.: Identifying structural flow defects in disordered solids using machine-learning methods. Phys. Rev. Lett. **114**, 108001 (2015)

El-Ganainy, R., Makris, K.G., Khajavikhan, M., Musslimani, Z.H., Rotter, S., Christodoulides, D.N.: Non-Hermitian physics and PT symmetry. Nat. Phys. **14**, 11–19 (2018)

Feng, L., El-Ganainy, R., Ge, L.: Non-Hermitian photonics based on parity–time symmetry. Nat. Photonics **11**, 752–762 (2017)

Hasan, M.Z., Kane, C.L.: Colloquium: topological insulators. Rev. Mod. Phys. **82**, 3045–3067 (2010)

Karimi, E., Schulz, S.A., De Leon, I., Qassim, H., Upham, J., Boyd, R.W.: Generating optical orbital angular momentum at visible wavelengths using a plasmonic metasurface. Light Sci. Appl. **3**, e167 (2014)

Khanikaev, A.B., Mousavi, S.H., Tse, W.K., Kargarian, M., MacDonald, A.H., Shvets, G.: Photonic topological insulators. Nat. Mater. **12**, 233–239 (2013). https://doi.org/10.1038/nmat3520

Krishnamoorthy, H.N., Jacob, Z., Narimanov, E., Kretzschmar, I., Menon, V.M.: Topological transitions in metamaterials. Science **336**, 205–209 (2012)

Lustig, E., Sharabi, Y., Segev, M.: Topological aspects of photonic time crystals. Optica **5**, 1390–1395 (2018)

Lustig, E., Weimann, S., Plotnik, Y., Lumer, Y., Bandres, M.A., Szameit, A. Segev, M.: Photonic topological insulator in synthetic dimensions. Nature **1** (2019)

Lu, L., Joannopoulos, J.D., Soljačić, M.: Topological photonics. Nat. Photonics **8**, 821–829 (2014)

Mirhosseini, M., Malik, M., Shi, Z., Boyd, R.W.: Efficient separation of the orbital angular momentum eigenstates of light. Nat. Commun. **4**, 2781 (2013)

Piao, X., Yu, S., Lee, M., Park, N.: Topological interface between anisotropic materials for transverse spinning of light fields. In: 2018 12th International Congress on Artificial Materials for Novel Wave Phenomena (Metamaterials), Espoo, pp. 314–315 (2018a)

Piao, X., Yu, S., Park, N.: Design of transverse spinning of light with globally unique handedness. Phys. Rev. Lett. **120**, 203901 (2018b)

Piao, X., Yu, S., Park, N.: Fano-resonant excitations of generalized optical spin waves. In: Kamenetskii, E., Sadreev, A., Miroshnichenko, A. (eds.), pp. 33–55. Springer International Publishing, Cham (2018c)

Rechtsman, M.C., Zeuner, J.M., Plotnik, Y., Lumer, Y., Podolsky, D., Dreisow, F., Nolte, S., Segev, M., Szameit, A.: Photonic Floquet topological insulators. Nature **496**, 196–200 (2013)

Shcherbakov, M.R., Vabishchevich, P.P., Shorokhov, A.S., Chong, K.E., Choi, D.-Y., Staude, I., Miroshnichenko, A.E., Neshev, D.N., Fedyanin, A.A., Kivshar, Y.S.: Ultrafast all-optical switching with magnetic resonances in nonlinear dielectric nanostructures. Nano Lett. **15**, 6985–6990 (2015)

Viciani, S., Lima, M., Bellini, M., Caruso, F.: Observation of noise-assisted transport in an all-optical cavity-based network. Phys. Rev. Lett. **115**, 083601 (2015)

Vitullo, D.L., Leary, C.C., Gregg, P., Smith, R.A., Reddy, D.V., Ramachandran, S., Raymer, M.G.: Observation of interaction of spin and intrinsic orbital angular momentum of light. Phys. Rev. Lett. **118**, 083601 (2017)

Xu, S., Wu, C.: Space-time crystal and space-time group. Phys. Rev. Lett. **120**, 096401 (2018)

Yu, Z., Fan, S.: Complete optical isolation created by indirect interband photonic transitions. Nat. Photonics **3**, 91–94 (2009). https://doi.org/10.1038/nphoton.2008.273

Yu, S., Koo, S., Park, N.: Coded output photonic A/D converter based on photonic crystal slow-light structures. Opt. Express **16**, 13752–13757 (2008a)

Yu, S., Koo, S., Piao, X., Park, N.: Application of slow-light photonic crystal structures for ultra-high speed all-optical analog-to-digital conversion. In: Microoptics Group (OSJ/JSAP), Tokyo. Paper D1 (2009). https://www.researchgate.net/publication/229007564_Application_of_slow-light_photonic_crystal_structures_for_ultra-high_speed_all-optical_analog-to-digital_conversion

Yu, S., Piao, X., Koo, S., Shin, J.H., Lee, S.H., Min, B., Park, N.: Mode junction photonics with a symmetry-breaking arrangement of mode-orthogonal heterostructures. Opt. Express **19**, 25500–25511 (2011). https://doi.org/10.1364/OE.19.025500

Yu, S., Piao, X., Park, N.: Slow-light dispersion properties of multiatomic multiband coupled-resonator optical waveguides. Phys. Rev. A **85**, 023823 (2012)

Yu, S., Piao, X., Hong, J., Park, N.: Progress toward high-Q perfect absorption: a Fano antilaser. Phys. Rev. A **92**, 011802 (2015)

Yu, S., Park, H.S., Piao, X., Min, B., Park, N.: Low-dimensional optical chirality in complex potentials. Optica **3**, 1025 (2016a). https://doi.org/10.1364/OPTICA.3.001025

Yu, S., Piao, X., Park, N.: Acceleration toward polarization singularity inspired by relativistic E × B drift. Sci. Rep. **6**, 37754 (2016b)

Yu, S., Piao, X., Cho, C., Park, N.: Phase manipulation of constant-intensity waves in disordered optical structures. In: 2018 12th International Congress on Artificial Materials for Novel Wave Phenomena (Metamaterials), Espoo, pp. 457–459 (2018a)

Yu, S., Piao, X., Park, N.: Bohmian photonics for independent control of the phase and amplitude of waves. Phys. Rev. Lett. **120**, 193902 (2018b)

Yu, S., Piao, X., Park, N.: Independent manipulation of amplitude and phase of light based on the de Broglie-Bohm viewpoint. In: 2018 12th International Congress on Artificial Materials for Novel Wave Phenomena (Metamaterials), Espoo, pp. 308–310 (2018c)

Yu, S., Piao, X., Park, N.: Designing non-Hermitian dynamics for conservative state evolution on the Bloch sphere. Phys. Rev. A **97**, 033805 (2018d)

Zhang, J., Hess, P., Kyprianidis, A., Becker, P., Lee, A., Smith, J., Pagano, G., Potirniche, I.-D., Potter, A.C., Vishwanath, A.: Observation of a discrete time crystal. Nature **543**, 217–220 (2017)

Index

© The Author(s), under exclusive license to Springer Nature Singapore Pte. Ltd. 2019   87
S. Yu et al., *Top-Down Design of Disordered Photonic Structures*,
SpringerBriefs in Physics, https://doi.org/10.1007/978-981-13-7527-9

Printed in the United States
By Bookmasters